STP 1462

Performance of Protective Clothing: Global Needs and Emerging Markets: 8th Symposium

Portia Yarborough and Cherilyn N. Nelson, Editors

ASTM Stock Number: STP1462

ASTM International
100 Barr Harbor Drive
PO Box C700
West Conshohocken, PA 19428-2959

Printed in the U.S.A.

ISBN: 0-8031-3488-6
ISSN: 1040-3035

Peer Review Policy

Each paper published in this volume was evaluated by two peer reviewers and at least one editor. The authors addressed all of the reviewers' comments to the satisfaction of both the technical editor(s) and the ASTM International Committee on Publications.

To make technical information available as quickly as possible, the peer-reviewed papers in this publication were prepared "camera-ready" as submitted by the authors.

The quality of the papers in this publication reflects not only the obvious efforts of the authors and the technical editor(s), but also the work of the peer reviewers. In keeping with long-standing publication practices, ASTM International maintains the anonymity of the peer reviewers. The ASTM International Committee on Publications acknowledges with appreciation their dedication and contribution of time and effort on behalf of ASTM International.

Printed in Ann Arbor, MI
September 2005

Foreword

Performance of Protective Clothing: Global Needs and Emerging Markets, Eighth Symposium was held in Tampa, Florida on 13–14 January 2004. ASTM International Committee F23 on Protective Clothing served as sponsor. Symposium chairpersons and co-editors of this publication were Portia Yarborough, Senior Research Chemist, E. I. du Pont de Nemours & Company, Inc., Richmond, VA; and Dr. Cherilyn N. Nelson, Vice President, Fibre and Knitting Technology, Ansell Healthcare Products, Inc., Seneca, SC.

Contents

Overview

This Special Technical Publication (STP) contains papers that address a range of topics related to the performance of protective clothing for government and industrial workers, first responders, and civilians. These papers were presented at the Eighth International Symposium on the Performance of Protective Clothing, held in Tampa, Florida on January 13–14, 2004. The theme of the eighth symposium sponsored by ASTM committee F23 was "Global Needs and Emerging Markets."

Advancements have been made in the field of protective clothing based on new fabric and garment technologies. This symposium was designed to expand the awareness of safety professionals, industrial hygienists, manufacturers, suppliers, and end users regarding the application of standards and performance specifications in the development of protective clothing. The symposium also provided a forum for the exchange of ideas on new and emerging industrial needs for protective clothing.

There were over one hundred registered participants from the U.S., Canada, Asia, and Europe. The attendees included research scientists, engineers, apparel and materials manufacturers, industrial safety professionals, and private industry consultants. ASTM committee F23 chairman, Jeffrey O. Stull, opened the symposium with a keynote address focused on global standards for protective clothing.

Oral presentations of the technical papers were arranged into six main topic areas:

- Nuclear, biological, and chemical protective apparel
- Thermal comfort
- Protective clothing for the food industry
- Predicting product performance
- Standards and test method development
- Assessment of thermal protection

A poster session and reception were also held to promote networking among the attendees.

The papers in this volume are separated into five sections. In the Nuclear, Biological, Chemical Protective Apparel section, information is shared on how to effectively evaluate material choices for chemical protective apparel using established test methods. The latest developments in the field of electronic textiles and their use in protective apparel assemblies for hazardous or combative environments are described.

Worker comfort as a critical aspect of protective clothing is detailed in the Thermal Comfort section. Improvements in testing and measurement of thermal and evaporative resistance as it relates to worker comfort are shown. Methods to accurately simulate and measure human perspiration using a thermal manikin are described. The results of a wear trial with personal cooling systems for first responders are presented. The study outlines the benefits of using cooling vests with apparel for protection from hazardous materials.

Technical papers in the Predicting Product Performance section outline various factors that influence the performance of protective apparel. For optimum worker benefit, protective garments should be designed to provide a high level of protection without limiting the workers' performance on the job. Nondestructive test methods for predicting the service life of the apparel are described.

The intent of the Standards and Test Method Development section is to share ways to improve existing test methods and to review data for establishing new test methods. Part of the focus of the standards committees' work is to ensure test methods remain relevant for the industry. As new materials and protective technologies are developed, it is important to ensure the standards and test methods are capable of accurately measuring and defining the performance of the product.

The section on Thermal Protective Assessment highlights modeling, manikin testing, and the performance of various clothing systems for providing thermal protection. Skin burn prediction techniques were investigated for assessing thermal hazards and risks. The results of a thermal characterization study using a manikin are presented. A comparison of the performance of clothing systems is made in order to develop a novel approach to soldier flame protection.

On behalf of ASTM, we thank all of the reviewers for their time and effort in reviewing the technical papers. The quality of this publication is reflected in the high professional standards of the reviewers. I wish to personally extend special thanks and recognition to the ASTM editorial staff and the staff managers for their hard work. We also thank the committee F23 members for their support and contributions to the symposium.

These proceedings contribute to the current body of knowledge relating to protective clothing performance. ASTM committee F23 will continue to provide forums to discuss current and future protective apparel trends along with real world application of existing standards. We look forward to your participation in future symposia.

Portia D. Yarborough
Senior Research Chemist
E. I. du Pont de Nemours & Company, Inc.
Richmond, Virginia
Symposium Co-Chairperson and Editor
June 2005

SECTION 1: NUCLEAR, BIOLOGICAL, CHEMICAL PROTECTIVE APPAREL

Journal of ASTM International, July/August 2004, Vol. 1, No. 7
Paper ID JAI12133
Available online at www.astm.org

Jeffrey O. Stull[1] *and Robert E. McManus*[2]

Evaluation of Three Chemical Protective Clothing Materials for Selected Performance Properties

ABSTRACT: A study was conducted to evaluate selected properties of three different chemical protective clothing materials that are being considered for use by Dupont's DeLisle Site in Pass Christian, Mississippi for titanium chloride and related chemical service. Two of the materials represented new product offerings based on unique PTFE/fabric (HSF) laminates, while the third was a PVC/Nomex material currently used in chemical protective clothing at the Dupont DeLisle Site. The two PTFE/fabric laminates were further distinguished by the film thickness and type of surface finish. A number of tests were performed to measure permeation resistance against specific chemicals of interest: physical property testing to assess material strength and specialized tests to determine how the materials hold up to different environmental challenges (high temperature, cold temperature, and flame exposure). The study demonstrated how a series of test methods established by Committee F23 can be used to evaluate the material choices effectively for a specific set of chemical and physical exposure conditions.

KEYWORDS: chemical protective clothing, PTFE/fabric laminates, permeation testing, physical property testing, titanium chloride.

Introduction

The purpose of this study was to evaluate selected properties of three different chemical protective clothing materials that are being considered for use by Dupont's DeLisle Site in Pass Christian, Mississippi for titanium chloride and related chemical service. Two of the materials represent new product offerings from W. L. Gore & Associates, while the third is a StaSafe material currently used in suits at the Dupont DeLisle Site. A number of tests were performed to measure permeation resistance against specific chemicals of interest: physical property testing to assess material strength and specialized tests to determine how the materials hold up to different environmental challenges (high temperature, cold temperature, and flame exposure). The results of the testing provide a comparison of material performance.

Background

The Dupont DeLisle Site is the second largest producer of titanium dioxide in the world. The chlorine process used to manufacture titanium dioxide employs significant amounts of chlorine and produces titanium tetrachloride as an intermediate chemical. Further processing of the chemical entails stages of alternately cooling and heating the chemicals used in the process [1]. The overall process creates a number of hazards, both chemical and physical, due to the various extremes of chemical conditions that may be encountered by chemical workers. In order to

Manuscript received 12 January 2004; accepted for publication 10 May 2004; published July/August 2004.
Presented at ASTM Symposium on Performance of Protective Clothing: Global Needs and Emerging Markets: 8th Symposium on 13 January 2004 in Tampa, FL.

[1] President, International Personnel Protection, Inc., 10907 Wareham Court, Austin, TX.
[2] Occupational Health Specialist, DuPont DeLisle Site, P. O. Box 430, Kiln-DeLisle Road, Pass Christian, MS.

protect workers during a multitude of different operations, the Dupont DeLisle Site sought to select a uniform set of protective clothing for all principle worker hazards.

Approach

The following properties were measured for each material:

- Permeation Resistance
- Tensile Strength
- Tear Resistance
- Burst Strength
- Flame Resistance
- Conductive Heat Resistance (Insulation)
- Conductive Cold Resistance (Insulation)
- Cold Temperature Performance (Embrittlement Resistance)

Table 1 provides the referenced test method and specific properties measured for each test. The specific test procedures used in this study are described in the Procedures section.

TABLE 1—*Selected test methods, type of performance assessed, and measurements used for the evaluation of selected material properties.*

Property	Test Method	Type of Performance Assessed	Measurements
Permeation Resistance	ASTM F 739	Evaluates the ability of the material to prevent the passage of the chemical through the material on a molecular level	Actual breakthrough time, normalized breakthrough time, permeation rate
Tensile Strength	ASTM D 5034	Evaluates the strength of the material when pulled in one direction	Breaking strength (both material directions)
Tear Resistance	ASTM D 5589	Measures the force to complete the tearing of a material once a tear has been established already	Tear force (both material directions)
Burst Strength	ASTM D 3787	Measures the force required to burst a material when force is perpendicular to material	Bursting force
Flame Resistance	ASTM F 1358	Measures ease of material ignition and ability of material to self-extinguish once flame is removed	Afterflame time, burn distance, burning behavior (observations)
Conductive Heat Resistance (Insulation)	ASTM F 1060	Measures heat transfer through material when contacted by a hot surface	Time to pain, time to second-degree burn
Conductive Cold Resistance (Insulation)	ASTM F 1060 (modified)	Measures heat transfer through material when contacted by a cold surface	Time to specific temperature (10, 0, and −10°C)
Cold Temperature Performance (Embrittlement Resistance)	ASTM D 2136 (modified)	Measures effects of cold exposure on material integrity	Condition following cold exposure and rapid bend

Materials and Sampling

Test Materials

The following three materials were evaluated in this study:

- W. L. Gore 20 μ, 3-layer HSF laminate without repellent finish (Material A)
- W. L. Gore 10 μ, 3-layer HSF laminate with repellent finish (Material B)
- StaSafe Gralite PVC-coated Nomex (Material C)

Specific details for each material are provided in Table 2.

TABLE 2—*Construction details for the test materials.*

Material	Construction	Thickness (mm)*	Weight (g/m²)**
10 μ HSF*** Laminate	Film laminated between woven and knit fabrics	0.64	329
20 μ HSF Laminate	Film laminated between woven and knit fabrics	0.64	336
Gralite	PVC coated on one side of Nomex™ twill fabric	0.79	685

* Thickness measured per ASTM D 1777.
** Unit area weight measured per ASTM D 3776.
***HSF™ is a trademark of W. L. Gore & Associates.

Sampling

Every attempt was made to provide statistically meaningful measurements of performance properties for the three chemical protective clothing materials. Because the two W. L. Gore materials are prototype fabrics, they were provided in large sheet samples. Two different sheets for the two W. L. Gore materials were obtained at different times, and the selection of specimens was randomized from each sheet. For the StaSafe Gralite material, a single sheet of material was obtained in addition to a full-body coverall. Again, specimen selection was randomized between the sheet of material and the coverall.

Procedures

Permeation Resistance

Permeation resistance testing was performed in accordance with ASTM F 739, Standard Test Method for Resistance of Protective Clothing Materials to Permeation by Liquids or Gases Under Conditions of Continuous Contact. In this test, a circular material specimen was placed in a permeation test cell dividing the cell into two separate chambers. The test chemical was introduced into one chamber while dry nitrogen or water was used in the other chamber as the collection medium.

- In the case of a dry nitrogen collection medium, the permeation test system was configured in an open-loop mode where fresh collection medium was poured through the collection chamber at a rate of 250 mL/min with the exit stream periodically sampled for the presence of test chemical.

- In the case where water was used as the collection medium, the permeation test system was configured in a closed-loop mode where the amount of collection medium stayed fixed and was sampled periodically during the test.

All tests were conducted for an 8 h exposure period. Table 3 shows the test parameters used for each of the seven test chemicals.

TABLE 3—*Test parameters for the measurement of material permeation resistance.*

Chemical	Test Configuration	Collection Medium	Sampling Interval	Analytical Technique
Ammonia (gas)	Closed loop	Water	15 min/30 min*	Conductivity
Chlorine (gas)	Open loop	Dry air	15 min/30 min*	Draeger Tube
Hydrogen Fluoride (gas)	Closed loop	Water	15 min/30 min*	Conductivity
Sulfuric Acid (96 %)	Closed loop	Water	15 min/30 min*	Conductivity
Titanium Tetrachloride	Closed loop	Water	15 min/30 min*	Conductivity
Toluene	Open loop	Dry nitrogen	6 min	Photoionization
Xylene (Mixed Isomers)	Open loop	Dry nitrogen	6 min	Photoionization

*Fifteen min for the first hour, and 30 min for every subsequent hour.

The actual breakthrough time was reported as the beginning time of the sampling interval in which chemical was first detected. The normalized breakthrough time was reported as the beginning time for the sampling interval where the rate of permeation was determined to be 0.1 $\mu g/cm^2/min$. Since the minimum detectable permeation rate for each chemical was lower than 0.1 $\mu g/cm^2/min$, the normalized breakthrough time was sometimes longer than the actual breakthrough time. The maximum permeation rate observed over the test period was also reported in units of $\mu g/cm^2/min$. If no breakthrough was detected during the test period, the breakthrough time was reported as greater than the length of the test period (8 h or 480 min), and the permeation rate was not reported.

Consistent with the test method, three total specimens were tested for each material and chemical. A fourth specimen was also tested as a blank and negative control.

Physical Properties

Tensile Strength—Tensile strength testing was performed in accordance with ASTM D 5034, Standard Test Method for Breaking Strength and Elongation of Textile Fabrics (Grab Test), using a constant rate-of-extension tensile test machine. In this test, a 100 mm × 200 mm material specimen was placed in the tensile testing machine, in which one end was held by a fixed grip (clamp), while the other end was held in a grip that was pulled at the uniform rate of 30 mm/min. A sensing device (load cell) on the moving grip measured the force as the specimen was pulled, registering the maximum force when the material specimen broke. Since the materials are produced in rolls and the material can have different strengths parallel to the direction of the roll (warp or machine direction) as compared to across the roll (fill or cross machine direction), the tensile or breaking strength was reported in the two directions. Consistent with the test method, five total specimens were tested in each material direction. The results in each direction were averaged and reported independently.

Tear Resistance—Tear resistance testing was performed in accordance with ASTM D 5587, Standard Test Method for Tearing Strength of Fabrics by Trapezoid Procedure. In this test, a 75 mm × 150 mm specimen was prepared by drawing a trapezoidal shape on the specimen with a notch cut along the short side of the trapezoid. The specimen was placed in the grips of a tensile tearing machine with the grips clamping the specimen along the drawn sides of the trapezoid. As with the tensile strength test (above), the material specimen was pulled at a rate of 12 in./min until it was completely torn apart. The average of the five highest measured forces was reported as the tearing strength. Consistent with the test method, five total specimens were tested in each material direction. The results in each direction were averaged and reported independently.

Burst Strength—Burst strength testing was performed in accordance with ASTM D 3787, Test Method for Bursting Strength of Textiles-Constant-Rate-of-Traverse (CRT) Ball Burst Test. In this test, a 125 mm diameter material specimen was clamped in a fixture that held the specimen in a horizontal position. This fixture was attached to the upper, movable arm of a tensile testing machine. A 25 mm steel ball affixed to the end of the rod was attached to the fixed arm of the tensile testing machine. The fixture was then moved at a rate of 30 mm/min, driving the ball through the material. The maximum force recorded as the ball ruptured the material was reported as the burst strength. Consistent with the test method, ten total specimens were tested.

Flame Resistance—Flame resistance testing was performed in accordance with ASTM F 1358, Standard Test Method for Effects of Flame Impingement on Materials Used in Protective Clothing Not Designated Primarily for Flame Resistance. In this test, a 75 mm × 405 mm material specimen was folded in half across its long axis using a 6.4 mm diameter metal rod. The metal rod was removed, and the folded specimen was placed in a holder, so that the folded edge protruded out from the bottom of the holder. The holder with the specimen was then hung from the suspension rack in a vertical flame chamber. The folded edge of the specimen was then exposed halfway into a methane-fueled flame adjusted to a 1½ in. height for a period of 3 s. Testing then proceeded using the following steps:

- If the specimen ignited (burning beyond 1 s after removal of the flame), the test was discontinued, and the continued burning time was reported as the afterflame time. The length of burn damage on the specimen was measured and reported as the burn distance.
- If the specimen did not ignite during the 3 s flame exposure, the flame was reapplied in an identical manner to the same specimen for an additional 12 s.
- If the sample then ignited during this subsequent exposure, the time the material specimen continued to burn following the removal of the flame was noted as afterflame time. The length of burn damage on the specimen was measured and reported as burn distance.

Observations of burning behavior, such as melting and dripping, were reported. Consistent with the test method, five total specimens were tested in each material direction. The results in each direction were averaged and reported independently.

Conductive Heat Resistance (Insulation)—Conductive heat resistance was measured in accordance with ASTM F 1060, Standard Test Method for Thermal Protective Performance of Materials for Protective Clothing for Hot Surface Contact. In this test method, a 100 mm × 150 mm specimen was placed on a hot plate heated to a temperature of 240 ± 3°C. This temperature was chosen arbitrarily as a potential hot liquid contact temperature for the protective clothing.

The test temperature was controlled through the use of an electrolytic copper plate on top of the hot plate that had a thermocouple mounted in the exact center of the plate. Immediately after the specimen was placed on the heated hot plate, a weighted sensor consisting of a copper calorimeter installed in an insulated material block and weighted with an aluminum block was placed on top of the specimen. The block created a pressure of 3.8 kPa on the sample. The temperature reading output of the sensor was collected on a data logger. The exposure was continued until the temperature rise exceeded 30°C. Consistent with the test method, three separate specimens were evaluated for each material.

The temperature output of the sensor was plotted versus time and compared to the Stoll curves for the prediction of time to pain sensation and second-degree burn injury (see example curve in Fig. 1). The intersection of the test material curve with the respective Stoll curve defined the skin threshold response.

Conductive Cold Resistance (Insulation)—Conductive cold resistance or heat loss through material specimens when exposed to cold temperature was measured as a function of time when the material system was placed in contact with a horizontally-oriented slab of dry ice under a pressure of 3.8 kPa in ambient conditions at 25°C. The same copper calorimeter assembly used in ASTM F 1060, described above, was used to measure heat flux through and temperature of the material specimens. The sensor's response (in °C) was recorded with a data logger. The resulting plot of temperature versus time was used to determine the time to reach specific "skin" (sensor surface) temperatures. Consistent with the test method, three separate specimens were evaluated for each material.

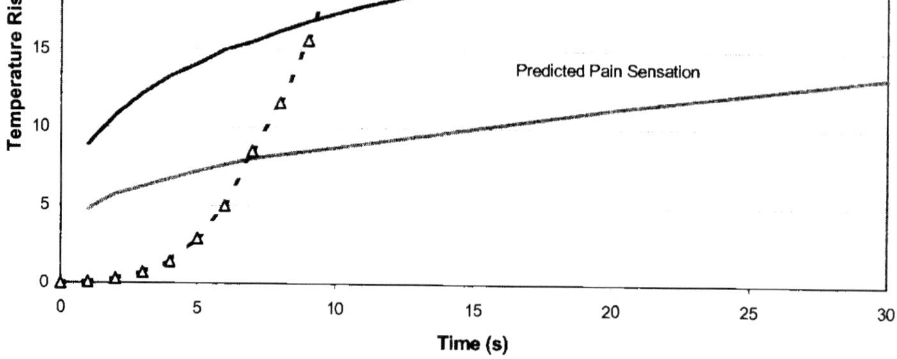

FIG. 1—*Prediction of time-to-pain and time-to-burn for measurement of conductive heat resistance.*

Unlike the Stoll curves used for measuring skin tolerance to high temperatures in terms of heat flux versus time, no similar information was available for establishing frostbite conditions. Three different temperatures (10°C, 0°C, and –10°C) were chosen for reporting the elapsed time from initial dry ice contact. These temperatures were chosen based on literature recommended values. Veghte [1] reported: "As skin temperatures fall below 10°C, pain can develop. As skin temperatures approach 0°C, numbness is likely to occur. As the extremities become accustomed to cold and skin temperatures approach freezing, the skin blood vessels adjust and open rhythmically (Lewis Effect). Warm blood then flows to the cold areas, and frostbite is prevented" and, "...where hands and feet were bare, tolerance time (defined as when any skin temperature reaches 0°C, was eight minutes." Another reference establishes tissue damage with contact at temperatures below 0°C [2].

Conductive Temperature Performance—Cold temperature performance, or embrittlement resistance, was measured using a modified form of ASTM D 2136, Standard Test Method for Coated Fabrics-Low-Temperature Bend Test, with the following modifications:

- Instead of a cold chamber exposure temperature where the chamber was set at a temperature of interest, a 25 mm × 100 mm material specimen was immersed in a Dewar flask containing liquid nitrogen (–210°C) using tongs for 10 min.
- After a 10 min immersion, the specimen was removed from the liquid nitrogen and inserted into the bending jig (Fig. 2), with the bending motion activated within 10 s of the specimen's insertion.
- The specimen was then removed and visually examined for evidence of cracking or other physical damage.

Consistent with the test method, three different specimens were evaluated for each material.

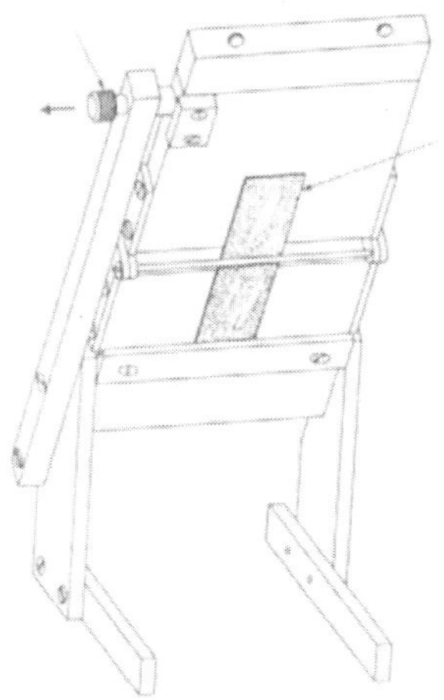

FIG. 2—*Bending jig for cold temperature embrittlement test.*

Results and Discussion

Permeation resistance test results are listed in Table 4, while all other testing results are summarized in Table 5.

Permeation Resistance

Table 4 lists the permeation test results for each material and chemical in terms of:

- Actual breakthrough time
- Normalized breakthrough time
- Maximum observed permeation rate

Based on these results, discrimination among the three materials was demonstrated for all chemicals but chlorine gas. In the case of chlorine gas, all three materials showed no breakthrough over the 8 h test period.

The 10 μ, 3-layer HSF laminate with repellent finish provided the best performance, with only one chemical (hydrogen fluoride) permeating the material within the 8 h test period. A second chemical, toluene, did show an actual permeation breakthrough time in one test cell, but the rate of permeation was less than the 0.1 μg/cm²/min used for the determination of normalized breakthrough time.

In general, the 10 μ, 3-layer HSF laminate with repellent finish outperformed the 20 μ, 3-layer HSF laminate without repellent finish, and the 20 μ, 3-layer HSF laminate without repellent finish outperformed the StaSafe PVC-coated Nomex.

- Significant differences between the 10 and 20 μ HSF materials were noted for ammonia, titanium tetrachloride, and toluene.
- The StaSafe Gralite material showed significant differences with the HSF materials for concentrated sulfuric acid, toluene, and xylene.

While there are no established industry requirements for permeation, guidelines for clothing selection against skin hazard chemicals can be found in Ref. [4]. In this standard, acceptability of the material is judged for providing a normalized breakthrough time of 1 h or more. Applying this guideline, the two 3-layer HSF laminates would be considered acceptable for all chemicals tested, except for hydrogen fluoride. The normalized breakthrough times for the 10 μ and 20 μ, 3-layer HSF materials were 45 and 30 min, respectively. If the 1 h breakthrough time criterion were applied to the StaSafe Gralite, then acceptable results would be obtained for ammonia, chlorine, and titanium tetrachloride.

TABLE 4—*Comparison of chemical resistance for selected protective clothing materials.*

Chemical	State	Measurement*	Material A	Material B	Material C
Ammonia	Gas	BT	> 480	370	380
		NBT	> 480	370	380
		PR	< 0.01	4.92	5.45
Chlorine	Gas	BT	> 480	> 480	> 480
		NBT	> 480	> 480	> 480
		PR	< 0.01	< 0.01	< 0.01
Hydrogen Fluoride	Gas	BT	45	30	25
		NBT	45	30	25
		PR	26.7	214.9	682.5
Sulfuric Acid, 96 %	Liquid	BT	> 480	> 480	30
		NBT	> 480	> 480	35
		PR	< 0.01	< 0.01	287
Titanium Tetrachloride	Liquid	BT	> 480	320	290
		NBT	> 480	360	440
		PR	< 0.01	0.35	0.17
Toluene	Liquid	BT	320**	188	7
		NBT	> 480	212	7
		PR	< 0.01	0.71	10.6
Xylene (Mixed Isomers)	Liquid	BT	> 480	> 480	4
		NBT	> 480	> 480	6
		PR	< 0.01	< 0.01	53.9

*Abbreviations: BT = Actual breakthrough time (min); NBT = Normalized breakthrough time (min); PR = Maximum permeation rate (μg/cm²/min). **Breakthrough was detected for only one of the three test cells.

TABLE 5—*Comparison of physical properties for selected protective clothing materials.*

Performance Property	Test Method	Measurement	Material A	Material B	Material C
Tensile Strength – Grab Method	ASTM D 5034	Breaking force (N)	1193 (warp) 939 (fill)	1104 (warp) 1442 (fill)	1437 (warp) 1104 (fill)
Burst Strength – Ball/Ring Method	ASTM D 3787	Burst force (N)	1255	1469	1558
Tear Resistance – Trapezoid Method	ASTM D 5587	Tearing force (N)	210 (warp) 185 (fill)	211 (warp) 258 (fill)	89 (warp) 60 (fill)
Flame Resistance – Folded Edge Method	ASTM F 1358	3 s ignition? (# ignited/# exposed)	Yes (5/5) (warp) Yes (5/5) (fill)	Yes (2/5) (warp) Yes (4/5) (fill)	No (0/0) (warp) No (0/0) (fill)
		Afterflame (s) for 3 s exposure	1.2 (warp) 1.3 (fill)	1.1 (warp)*** 1.1 (fill)***	Not applicable
		12 s ignition? (# ignited/# exposed)	Not applicable	No (3/5) (warp) No (1/5) (fill)	Yes (3/5) (warp) Yes (5/5) (fill)
		Afterflame (s) for 12 s exposure	Not applicable	Not applicable	1.0 (warp)^ 1.0 (fill)
		Burn distance (mm)	2.5 (warp) < 2.5 (fill)	13 (warp) 6 (fill)	10 (warp) 13 (fill)
		Specimen dripped?	No	No	No
		Specimen melted?	No	No	No
Hot Contact Insulation	ASTM F 1060 at 240°C	Time to pain (s)	7.1	7.2	4.6
		Time to burn (s)	8.9	9.1	5.6
Cold Temperature Insulation	ASTM F 1060 (modified)*	Time to 10°C (s)	21	19	13
		Time to 0°C (s)	46	43	31
		Time to –10°C (s)	80	77	57
Cold Temperature Embrittlement	ASTM D 2136 (modified)**	Observation of fracture/cracking	No	No	Yes

* Cold surface based on dry ice with measurement of time to 0°C.
** Based on 10 min immersion in liquid nitrogen.
*** Average applies only to specimens igniting in 3 s flame exposure.
^ Average applies only to specimens igniting in 12 s flame exposure.

Physical Properties

Strength Properties—The physical property test results in Table 5 showed some differences in the three materials:

1. The fewest differences were found for tensile strength testing.
2. The StaSafe PVC-coated Nomex had the highest burst strength.

The HSF laminates had comparable tear resistance, which was much greater than the StaSafe Gralite. There were some differences apparent in the average test results when examined by material direction. Since the direction of the received materials was not marked, a direction had to be assumed. This may explain why there was no consistency between materials for which direction yielded higher average results.

To provide perspective for these test results, a comparison was made with the minimum requirements for physical properties in the following standards (this comparison is shown in Table 6):

- NFPA 1991, Standard for Vapor-Protective Ensembles for Hazardous Material Emergencies (2000 Edition) [4]
- NFPA 1992, Standard for Liquid Splash-Protective Ensembles and Clothing for Hazardous Material Emergencies (2000 Edition) [5]
- ANSI/ISEA 103, American National Standard for Classification and Performance Requirements for Chemical Protective Clothing (proposed) [6]

TABLE 6—*Comparison of material physical properties with relevant standards.*

Property	Material	Result*	Minimum Requirement		
			NFPA 1991	NFPA 1992	ISEA 103**
Tensile Strength	Material A	1193/939	No requirement	No requirement	No requirement
	Material B	1104/1442			
	Material C	1437/1104			
Burst Strength	Material A	1255	> 45	> 30	Class 6 > 1600 Class 5 > 800
	Material B	1469			
	Material C	1558			
Tear Resistance	Material A	210/185	No requirement	No requirement	Class 6 > 200 Class 5 > 150 Class 3 > 50
	Material B	211/258			
	Material C	89/60			

* All results are in Newtons. When two results are reported, the first result is for the machine direction, and the second result is for the cross-machine direction.
** Level is based on classification scale for property.

Flame Resistance—Flame resistance test results are provided in Table 5. These results are provided in seven parts:

1. The first result indicates whether or not any of the specimens ignited during the 3 s exposure and provides the number of ignited specimens out of the total five specimens tested. According to ASTM F 1358, ignition occurs when the afterflame time is at least 1 s long.
2. The second result provides the average afterflame time for each direction following the 3 s exposure.
3. The third result indicates whether or not any of the specimens ignited during the subsequent 12 s exposure and provides the number of ignited specimens out of the total specimens tested. If some specimens ignited during the 3 s exposure, the total number of specimens tested was less than five.
4. The fourth result provides the average afterflame time for each direction for the group of specimens that was subjected to the 12 s exposure.
5. The fifth result presents the average burn distance by material direction for all specimens tested.
6. The sixth and seventh results indicate whether the specimens melted or dripped.

Mixed results were found during flame resistance testing:

- The StaSafe Gralite showed no ignition during the initial 3 s exposure period and then relatively short afterflame times and short burn distances for the 12 s exposure period.
- The 10 μ, 3-layer HSF laminate exhibited ignition for all specimens during the 3 s exposure period, but the afterflame times were only slightly over the prescribed 1 s threshold (for not applying the 12 s exposure), and the burn distances were very small.
- The 20 μ, 3-layer HSF laminate had some specimens ignite during the 3 s exposure, while other specimens did not ignite.
- None of the materials showed any melting or dripping.

NFPA 1991 sets minimum baseline criteria for the flame resistance of encapsulating chemical protective suits. These criteria specify that the material should not ignite during the initial 3 s exposure period, and that the resulting afterflame time should be less than 10 s and the burn distance should be less than 100 mm. Comparing the tested materials against these criteria, only the StaSafe Gralite would comply.

Thermal Insulation and Embrittlement—The three materials were tested for both heat and cold conductive resistance as well as embrittlement resistance to cold temperature contact. The results of this testing appear in Table 5.

The conductive heat resistance testing was conducted at a contact temperature of 240°C. The temperature rise for all specimens was relatively instantaneous. Both 3-layer HSF laminates provided similar performance, which was slightly better than the StaSafe Gralite. It is likely that the fabric/film/fabric construction of the material was responsible for this advantage, even though the Gralite material is significantly heavier. Denser materials tend to transmit heat at a much faster rate, and contact with the coating side of the material creates a significant pathway for heat transfer through the specimen. Nevertheless, the predicted times to both pain sensation and second-degree burn are relatively short for all three materials (predicted burn times were all less than 10 s). Therefore, if heat protection was needed, supplemental protective clothing would be recommended.

In testing the cold temperature insulation of the three materials, the time required for reaching specific temperatures that are considered significant for skin pain and damage (from frostbite) were examined. Dry ice was used as a contact surface because it establishes an exposure temperature close to the boiling point of many liquefied gases. The trends in material performance were similar for the conductive heat resistance testing; however, the threshold times were substantially longer. It is therefore likely that these materials could provide cold protection for a limited period of time without injury, though extended contact would warrant additional insulation.

The cold temperature embrittlement tests showed passing performance of the HSF materials when exposed to liquid nitrogen. These materials exhibited only mild stiffening. However, the StaSafe Gralite material showed evidence of cracking when rapidly bent after being immersed in liquid nitrogen.

Conclusions

A range of performance properties has been evaluated for each of the three protective clothing materials. The results of this testing distinguish specific performance advantages of each material when compared with one another. For example, the 10 μ, 3-layer HSF material with repellent finish provided the longest breakthrough times and smallest permeation rates for the battery of test chemicals (when measured). This material utilized a film that has been optimized for chemical permeation resistance compared with the 20 μ, 3-layer HSF material. Physical property differences were less significant between materials. The StaSafe Gralite offered the greatest flame resistance. The two 3-layer HSF materials provided slightly better conductive resistance against both hot and cold surface exposures and resisted the effects of embrittlement when compared to the StaSafe Gralite material.

While the evaluations conducted as part of this study connote a relatively thorough investigation of chemical protective clothing performance against different hazards, the number and type of evaluations conducted do not include all appropriate tests and may not represent all working conditions that may be encountered during use in chemical protective clothing. Furthermore, aspects related to the use of these materials in specific clothing designs can significantly affect the overall protection of the protective clothing.

References

[1] *Dupont Literature on the Manufacture of Titanium Dioxide*, 2003.

[2] Veghte, J. H., "Design Criteria for Fire Fighters' Protective Clothing," Janesville Apparel, Lion Uniform, Dayton, Ohio, 1981, p. 13.

[3] Woodson, W. E., *Human Factors Design Handbook*, Van Nostrand, New York, 1992, p. 812.

[4] NFPA 1991, "Standard for Vapor-Protective Ensembles for Hazardous Material Emergencies," National Fire Protection Association, 1 Batterymarch Park, Quincy, MA, 2000.

[5] NFPA 1992, "Standard for Liquid Splash-Protective Ensembles and Clothing for Hazardous Material Emergencies," National Fire Protection Association, 1 Batterymarch Park, Quincy, MA, 2000.

[6] ANSI/ISEA 103, "American National Standard for Classification and Performance Requirements for Chemical Protective Clothing (Proposed)," International Safety Equipment Association, Arlington, VA.

Journal of ASTM International, July/August 2005, Vol. 2, No. 7
Paper ID JAI12104
Available online at www.astm.org

Carole A. Winterhalter,[1] Justyna Teverovsky,[2] Patricia Wilson,[3] Jeremiah Slade,[4] Brian Farell,[5] Wendy Horowitz,[6] and Edward Tierney[7]

Development of Electronic Textiles to Transport Data and Power in Future U.S. Military Protective Clothing Systems

ABSTRACT: The focus of this paper is on the development of textile-based wearable electronics that can be integrated into military protective clothing. A materials and manufacturing survey was conducted to determine the best performing and most durable materials to withstand the rigors of textile manufacturing and potential military use. Narrow woven technology was selected as the most appropriate manufacturing method. A working wearable narrow fabric version of the Universal Serial Bus (USB), as well as a radiating conductor, were successfully developed and fabricated. Military products developed include components of a personal area network providing data and power transport and a body borne antenna integrated into a load-bearing vest.

KEYWORDS: electronic textiles, textile data and power bus, personal area network, wearable antenna, military, army, soldier

Introduction

The need for real time information technology on the battlefield has been well documented in programs such as Land Warrior, Scorpion, and Future Force Warrior (FFW). Military combat clothing materials are passive, and the ability to integrate electronics into textiles provides the potential to achieve revolutionary improvements in performance and the realization of capabilities never before imagined on the battlefield. Electronic devices are being miniaturized for personal use both in the commercial and military sectors. Materials and methods are being investigated to facilitate the integration of these electronics into textiles. While there is an increasing number of commercial electro-textile products on the market, such as the Sensatex Smart Shirt, Vivometrics' Lifeshirt, and the Malden Mill's Heat Blanket, this effort focuses on the development of working prototypes specifically designed to support rugged military use in future protective clothing systems.

Two selected military applications are textile-based cables to support a wearable electronic network providing data and power transport, and a textile-based double loop antenna that can be integrated into protective clothing. Projected capability enhancements on the battlefield

Manuscript received 26 November 2003; accepted for publication 22 December 2004; published July 2005.
Presented at ASTM Symposium on Performance of Protective Clothing: Global Needs and Emerging Markets: 8th Symposium on 13 January 2004 in Tampa, FL.
[1] Textile Technologist, U.S. Army Research, Development and Engineering Command, Natick, MA 01760-5019.
[2] Senior Engineer, Foster-Miller, Inc., Waltham, MA 02451-1196.
[3] Principal Scientist, Foster-Miller, Inc., Waltham, MA 02451-1196.
[4] Project Engineer, Foster-Miller, Inc., Waltham, MA 02451-1196.
[5] Manager of Electronic and Biomedical Material Systems Division, Foster-Miller, Inc., Waltham, MA 02451-1196.
[6] Director of Marketing, Offray Specialty Narrow Fabrics, Chester, NJ 07930-0421.
[7] Development Engineer, Offray Specialty Narrow Fabrics, Chester, NJ 07930-0421.

provided by these products include an individual soldier network to process and use information in real time, resulting in increased situational awareness and enhanced communications.

The objectives of this effort are to investigate the feasibility of integrating conductive materials and components into textiles and to develop working prototypes of electronic textiles for potential future military use. A materials and textile manufacturing survey [1] was conducted to determine the best performing and most durable materials to withstand the rigors of textile manufacturing and military use. A variety of commercial materials were investigated, tested, and evaluated including conductive polymers, metallic fibers, novelty embroidery materials, and materials not commonly used in textile manufacture, such as tinsel wire and optical fiber. Copper wire of 32 American Wire Gauge (AWG) was used as a benchmark for conductivity and has a value of 186 siemens per centimeter. None of the fiber-forming polymers investigated demonstrated conductivities approaching copper wire. Other polymers such as polyaniline and polypyrrole are conductive, but the level did not approach that of copper, and the mechanical properties of the polymers were known to be poor. Metallic fibers such as Aracon® [2], which is metal coated Kevlar, and X-static® [3], which is a silver coated nylon, are both available in broad product ranges, but did not approach the desired conductivity of copper. Embroidery materials, such as metalized Mylar wrapped around core fibers, demonstrated relatively high resistance and poor mechanical properties. Tinsel wire, commonly used in the telecommunications industry, is composed of a bundle of core fibers, usually polyester or Kevlar, which is double wrapped with metallic foil. The typical resistance of an individual tinsel wire is 0.015 ohms per centimeter. Typical resistance values of copper wire are listed in Table 1.

TABLE 1—*Benchmark conductors.*

Solid Copper Wire, AWG	Resistance, ohm/cm	Diameter, mm
28	0.0021	0.32
30	0.0034	0.25
36	0.014	0.12

Due to the lack of highly conductive fiber-based materials, traditional conductors such as copper wire and tinsel wire were used in this development effort.

Glass and plastic optical fiber can also be used to transport data in electro-textile bus structures. Optical transport of data can take place at rates in excess of 1 Gb/s. Glass has been optimized for use in the telecommunications industry; however, their characteristic large bandwidths and operational distances are not necessary for use in protective clothing. Plastic optical fiber was selected because distance requirements are moderate (<5m), and fine glass fragments may complicate battlefield wounds.

Other factors to be considered when selecting plastic optical fiber are:

- High bandwidth, reduces the number of data transmission lines
- Electromagnetic interference shielding is not required
- Resistant to nicks, can be melted together in the field to repair
- Bend radii are small (<1 mm)
- Bend radii less than 25 mm result in transmission loss
- Compatible connector systems not well developed

Traditional textile manufacturing methods were investigated to determine if fine gauge wire as well as tinsel wire and optical fiber could be integrated into a manufactured fabric. Broad loom and narrow woven fabrics, knits, and embroidery methods were reviewed. A mass producible, flexible, washable, and low cost wearable textile fabric was desired. While conductive materials can be easily integrated into broadloom woven and warp knit fabrics, the problem is that the conductive path is lost when pattern pieces are cut and joined together. Narrow fabrics need only be terminated at the ends. Narrow woven fabric can be used to create a continuous network on or within a garment or clothing system without seam discontinuities and could be applied to any garment. In addition, narrow fabric manufacturing technology showed promise because the double plain weave construction is uniquely compatible with our interests to integrate both conductive and optical materials into the base fabric. The double plain weave, which is one of the most common types of weaves used for medium weight industrial webbings, consists of two layers of fabrics, one above the other, each woven with a separate warp and filling system. Binder yarns are commonly used to stitch the top and bottom layers together. The binders divide the space between the two layers into warp-wise tubes, providing space for stuffer components. Stuffer components do not interlace with any yarns, lay flat, and are not visible from the face of the cloth. Traditionally, stuffer yarns are used to increase thickness and strength, or provide a rib effect. In our experiment we used conductive and optical fibers as the stuffer components. In addition, narrow woven double cloths usually have one woven selvage and one knitted selvage. The knitted selvage incorporates a catch cord that can be used to facilitate unraveling of the narrow fabric for subsequent connection purposes.

Materials and Methods

While copper wire is not generally thought of as being flexible or textile compatible, it is compatible with all wire and data transmission standards investigated in this development effort. Solid copper wire is stiff, and it was recognized that by constructing wire using a larger number of high gauge, i.e., small diameter, fine filaments, it might be possible to increase flexibility. However, when 22 AWG power wire and 28 AWG shielded twisted pair products were compared on a cost basis, it was found that as the filament gauge increased, so did the cost. Figure 1 shows the relationship between the cost per 1000 feet of wire and filament gauge and demonstrates that copper wire filaments greater than 44–46 AWG in size are not cost effective.

Electro-Textile Data and Power Transmission Device

The standard Universal Serial Bus (USB) 2.0 protocol was selected to prove the technical feasibility of manufacturing textile-based components of a point-to-point wearable personal area network because it is a cable bus that supports data exchange between a host computer and a broad range of peripherals, and it is commonly used and well understood. A typical USB cable (Fig. 2) is composed of a non-twisted power pair, twisted data pair, a drain wire, and various shielding.

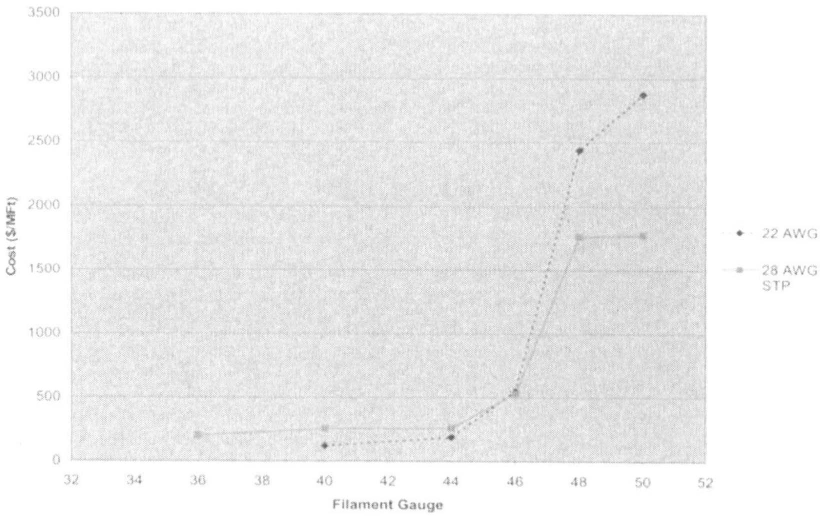

FIG. 1—*Unit cost of stranded wires as a function of filament gauge for 22 AWG Power Wire and 28 AWG Shielded Twisted Pair.*

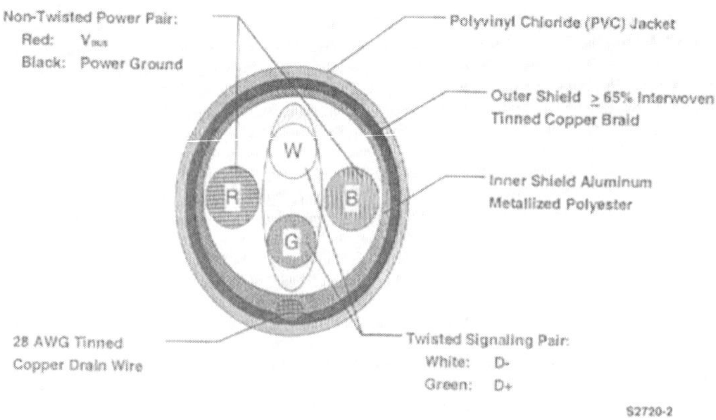

FIG. 2—*Cross-section of a standard USB 2.0 cable.*

A textile prototype was manufactured in accordance with the industry standard USB 2.0 specifications [4], and the following materials were used:

- Data Transmission Medium: 28 AWG twisted copper pair, PVC insulated, wrapped with aluminum Mylar foil for EMI shielding
- Power Wires: 20 AWG stranded, tinned copper with PVC insulation
- Drain: 28 AWG copper wire in contact with data medium
- Warp and Filling Yarn: Filament Nylon

The components selected for use in the narrow fabric version were similar to those used in the concentric design, but they were laid out in a flat arrangement (Fig. 3). The double plain weave construction was utilized, except the traditional stuffer components were substituted for electronic components. The gray stripe down the middle identifies the location of the electronic components to alert garment manufacturers to avoid stitching in this area.

FIG. 3—*Textile based USB. From top, white power wire, drain wire, shielded twisted signaling (data) pair, black power wire.*

Radiating Conductor

Narrow fabric manufacturing technology was also used to develop a radiating conductor for body borne antenna applications. A double plain weave fabric (Fig. 4) was manufactured with binder and stuffer components. Four, one-quarter-inch wide parallel conductive stripes made from tinsel warp yarns are separated by one-eighth inch wide insulators of nylon filament warp yarn. The optical fibers, which have a low bending radius, were integrated into the structure as the stuffer components. Nylon filament yarn was used for the binder and filling components. The textile based radiating conductor was waterproofed by dip coating in a thinned thermoplastic resin.

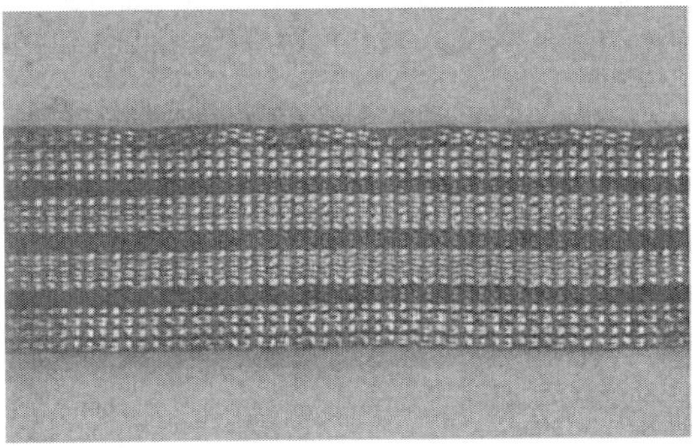

FIG. 4—*Textile-based radiating conductor.*

Results, Discussion, and Military Applications

The prototype electro-textile cable was tested and evaluated against a variety of tests including Group 6 Signal Integrity tests and shielding effectiveness for the USB 2.0 (Table 2) in accordance with industry standards [5,6].

TABLE 2—*Group 6 signal integrity test results.*

Cable Length	Cable Impedance (full speed)	Propagation Delay	Propagation Delay Skew	Attenuation (full speed)
5.0m	Pass	Pass	Pass	Fail
4.0m	Pass	Pass	Pass	Fail
3.0m	Pass	Pass	Pass	Pass
2.5m	Pass	Pass	Pass	Pass
2.0m	Pass	Pass	Pass	Pass
1.0m	Pass	Pass	Pass	Pass
0.5m	Pass	Pass	Pass	Pass

While the USB 2.0 tests are not intended for body worn applications, we selected those that would be appropriate for soldier applications. Overall, the cables passed all body worn related requirements. Impedance is a parameter that determines a cable's ability to pass energy efficiently. The specification requires that the impedance register between 76.5 and 103.4 ohms. All cables passed this test. Propagation delay is the time required for digital signal to travel from the input of a cable to the output. To pass the low speed spec, the propagation delay must be no more than 5.2 ns/m. The 0.5–4 m cables meet this requirement. The 5.0 cable is at the threshold with an average speed of 26.03 ns and a requirement of 26.0 ns. The propagation delay skew tests verify that the signal of both the D^+ and D^- lines arrive at the receiver at the same time. The standards require a differential of less than 100 ps. All cables passed this test. Attenuation is a measure of signal loss in a cable. Attenuation tests are only required for high or full speed cables and were performed to ensure that adequate signal strength is presented to the receiver to maintain low error rates. The USB cables failed at long lengths. For the purposes of the soldier, however, only short cable lengths will be required. The cables pass attenuation tests at lengths of 3.0 m and shorter. Shielding effectiveness tests were performed to determine the electromagnetic emission profile and susceptibility of the cable. The minimum attenuation required to pass the test is 20 dB. Testing occurred at 13 frequencies between 30 MHz and 1 GHz. All of the USB cables achieved the 20 dB mark at all frequencies, with the lowest being 35.9 dB.

To measure durability-related performance, the electro-textile cable was tested and evaluated after abrasion (4000 cycles [7]) and cyclic loading (0–250 pound load profile for 40 000 cycles). The electro-textile cables maintained acceptable performance levels after being subjected to abrasion and cyclic loading [8,9].

Future Force Warrior (FFW) is the future protective clothing system under development for soldiers of the U.S. Army. Components of a personal area network were developed utilizing the technology described above and integrated into the FFW prototype system to serve as the soldiers' electronic backbone. The textile-based power and data buses form a point-to-point network when hard wired to the central processing unit (CPU) (Fig. 5.). The CPU is designed to be worn in the small of the soldier's back. Power from the two waist-worn battery packs, one on each side, is transported to the CPU via the two side electro-textile cables. The three textile-based cables at the bottom of the CPU wrap under it, then between layers within the FFW system, emerging at the back of the neck to transport data and power to and from the wrist display, drink-o-meter, and headgear.

FIG. 5—*FFW CPU with attached electro-textile cables.*

The narrow fabric technology and conductive materials were used to develop a body worn VHF antenna. Maintaining communications on the battlefield is critical to coordinate and control units and firepower. Antennas that protrude from the soldier's silhouette easily identify radio operators. Antennas are easily broken by tree branches and bushes, limit soldier mobility, and are relatively inefficient radiators. A rigid Merenda double-loop antenna [10,11,12,13] was transformed into a wearable textile based antenna that is compatible with the Single Channel Ground and Airborne Radio System (SINCGARS) functioning in the 30–88 MHz range. The radiating conductors are attached to the antenna's high efficiency electronic switching components that increase the antenna's wideband efficiency. The gaps in the conductors (Fig. 6) indicate where the switching modules will be placed. The textile-based antenna was integrated into the Modular Lightweight Load Carrying Equipment (MOLLE) vest (Fig. 7). In this configuration, the body borne antenna has advantages over the standard 30-in. whip antenna in that it is body conformal and visually covert, not compromising the soldier's silhouette. RF hazard and safety testing was conducted by the U.S. Army Communications and Electronics Command [14] at the U.S Army's Electronic Proving Ground to measure Permissible Exposure Levels (PEL) inside and around the antenna vest. These tests indicated that the antenna vest met all PEL standards across the 30–88 MHz band. Performance testing was conducted by BAE Systems Aerospace Inc. [15], with a SINGCARS radio both on and off the human body. Testing was conducted on a human volunteer while standing, kneeling, and crouching close to the ground to measure its omni-directional performance. Test results (Fig. 8) indicated that it is possible to achieve communications ranges compatible with the SINCGARS radio and standard issue whip antenna.

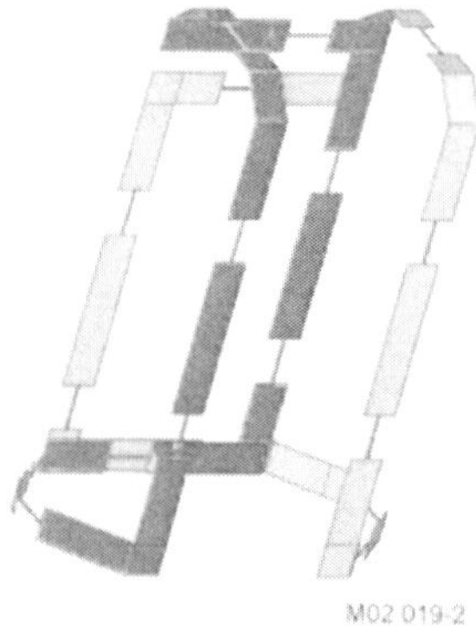

FIG. 6—*Textile components in Merenda Double Loop Antenna configuration.*

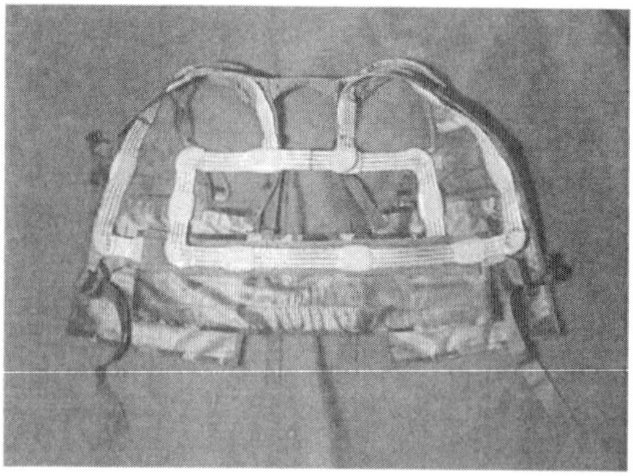

FIG. 7—*Double Loop Antenna integrated into the MOLLE vest.*

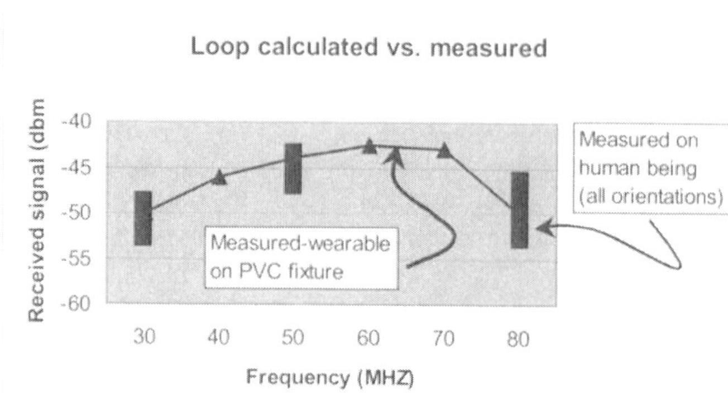

FIG. 8—*Radiation levels on human and PVC fixture.*

Conclusions

Electronic textiles were manufactured using commercially available materials and traditional textile manufacturing methods. Conductive materials such as copper wire, tinsel wire, and optical fiber were integrated into double plain weave narrow fabrics. Stranded copper wire filaments less than 44–46 AWG in size were found to be the most cost effective. Narrow woven electro-textile cables may be used to create a continuous network on or within a military garment or clothing system without seam discontinuities. A double plain weave narrow fabric with integrated conductive components was manufactured that met selected test requirements of the U.S.B. 2.0 initially and after exposure to abrasion and cyclic loading. A radiating narrow fabric with integrated conductive and optical materials was manufactured that demonstrated communications ranges compatible with military communications equipment.

References

[1] Wilson, P., Carey, C., Farrell, B., Gassner, J., Haugsjaa, P., Pan, M., et al., *Electro-Optic Fabrics for the Warrior of the 21st Century, Natick/TR-99/030L*, U.S. Army Natick Soldier Systems Center, Natick, MA, 1999, Defense Technical Information Center, Ft. Belvoir, VA, Online (URL: http://www.disq.mil/main/dtic.html).
[2] Aracon is trademark of E.I. duPont de Nemours & Company, Wilmington, DE.
[3] X-static is a trademark of Noble Fiber Technologies, Clarks Summit, PA.
[4] USB Implementers Forum, Universal Serial Bus Specification Revision 2.0, 2002, Online (URL: http://wwwusb.org.html/).
[5] USB Implementers Forum, Universal Serial Bus Cables and Connectors Class Document, Revision 2.0RC1, October 31, 2002, Online (URL: http://www.usb.org.html/).
[6] Electrical Industries Association, Electrical Connector/Socket Test Procedures Including Environmental Classifications, Arlington, VA, Online (URL: http://www.eia.org.html/).
[7] U.S. Federal Test Standard 191, Test Method 5309 - Abrasion Resistance of Textile Webbing, Online (URL: http://www.disq.mil/main/dtic.html).
[8] Slade, J., Wilson, P., Farrell, B., Teverosky, J., Thomson, D., Bowman, J., et al., "Mechanical Testing of Electrotextile Cables and Connectors," *Electronics on Unconventional Substrates - Electrotextiles and Giant-Area Flexible Circuits*, Materials Research Society, Shur, M., Wilson, P., Urban, D., Eds., Warrendale, PA, 2003.
[9] Slade, J., Wilson, P., Farrell, B., Teverovsky, J., Thomson, D., Bowman, J., et al., "Improving Electrotextile Wearability Using Stiffness Testing Methods," *Electronics on Unconventional Substrates - Electrotextiles and Giant - Area Flexible Circuits*, Materials Research Society, Shur, M., Wilson, P., Urban, D., Eds., Warrendale, PA, 2003.
[10] U.S. Patent "Synthesizer Radiating Systems and Methods," 5, 402,133, issued March 28, 1995, United States Patent and Trademark Office, Arlington, VA, Online (URL: http://www.uspto.gov/index.html).
[11] U.S. Patent, "Radiation Synthesizer Systems and Methods," issued May 8, 2001, United States Patent and Trademark Office, Arlington, VA, Online (URL: http://www.uspto.gov/index.html).
[12] U.S. Patent, "Radiation Synthesizer Feed Configurations," issued Aug 12, 2003, United States Patent and Trademark Office, Arlington, VA, Online (URL: http://www.uspto.gov/index.html).
[13] U.S. Patent, "Radiation Synthesizer Receive and Transmit Systems," issued Sep 2, 2003, United States Patent and Trademark Office, Arlington, VA, Online (URL: http://www.uspto.gov/index.html).
[14] Space and Terrestrial Communications Directorate.
[15] BAE Systems Aerospace, Inc., Advanced Systems, One Hazeltine Way, Greenlawn, NY.

SECTION 2: THERMAL COMFORT

Journal of ASTM International, April 2005, Vol. 2, No. 4
Paper ID JAI12109
Available online at www.astm.org

Mark G. M. Richards[1] *and Elizabeth A. McCullough*[2]

Revised Interlaboratory Study of Sweating Thermal Manikins Including Results from the Sweating Agile Thermal Manikin

ABSTRACT: Recently thermal and evaporative resistances of five clothing ensembles were measured using thermal manikins at six laboratories. For evaporative resistances, the manikin surfaces were wetted prior to or constantly during measurement. In this paper, previous resistance values are combined with results from the Sweating Agile Thermal Manikin of EMPA. For all ensembles, reproducibility of the intrinsic thermal resistance between labs improved by 4–9 % after including EMPA results and a further 16–39 % for four ensembles after excluding results from another lab. The intrinsic evaporative resistance reproducibility for three ensembles improved by 7–9 % after including EMPA and by 7–29 % excluding the other lab. In spite of these improvements, differences in manikin design, measurement, and calculation methodologies still account for the large reproducibilities ranging from 22–46 % and from 41–138 % for the intrinsic thermal and evaporative resistances, respectively. Presently, ASTM is drafting a new standard for sweating thermal manikins. Standardization and methodological improvement, including proposals presented here, should help to reduce these reproducibilities in future.

KEYWORDS: thermal manikin, sweating, evaporative resistance, clothing insulation, protective clothing

Introduction

Heat loss and thermal comfort of the human body are dependent on many factors, including the clothing being worn. The evaporation of sweat is an important means of body heat loss when a person is physically active or in a hot environment. Consequently, many body heat balance models and thermal comfort indexes require values of the thermal insulation and evaporative resistance of clothing. For flat textile layers, methods such as ISO 11092 "Textiles – Physiological effects – Measurement of thermal and water-vapour resistance under steady-state conditions" are used to determine these values. However, to determine these values for manufactured clothing ensembles, life-sized thermal manikins must be used.

Over the last 60 years, thermal manikins have been developed to measure the thermal insulation of clothing including several standards relating to manikin design and use [1]. Advances in design include the number of heated body segments and the addition of mechanisms to achieve automated repetitive body movements. Although the majority of manikins to date are adult-sized, recently a number of child and baby manikins have been developed (e.g., [2,3]). Over the last 15 years, a few thermal manikins have been used to measure the evaporative resistance of clothing by wetting the manikin's surface. Some of these wettable thermal manikins have an integrated sweating system, which enables a constant sweat rate to be defined and

Manuscript received 5 January 2004; accepted for publication 14 September 2004; published April 2005. Presented at ASTM Symposium on Performance of Protective Clothing: Global Needs and Emerging Markets: 8th Symposium on 13 January 2004 in Tampa, FL.
[1]Research Scientist, EMPA Swiss Federal Laboratories for Material Testing and Research, Lerchenfeldstrasse 5, CH-9014 St. Gallen, Switzerland.
[2]Professor and Co-Director, Institute for Environmental Research, Kansas State University, Manhattan, KS U.S.A.

maintained over a period of time, enabling steady-state conditions to be reached for a given sweat rate.

KSU Interlaboratory Study

An interlaboratory study of adult-sized sweating thermal manikins from six different laboratories was completed by Kansas State University (KSU), USA, in 2002 [4]. The laboratories which took part are: KSU, North Carolina State University (NCSU) USA, Navy Clothing and Textile Research Facility (NCTRF) USA, Centre of Human Sciences, QinetiQ Ltd. (QinetiQ) UK, Technical Research Centre of Finland (VTT), which has now moved to Tampere University of Technology, and Bunka Women's University (BWU) in Japan.

That study revealed a large spread of evaporative resistance results which can be attributed to the large differences in apparatus (manikin and climatic chamber), experimental procedures, and methods of calculation used. Details of the various manikins with their different sweating systems and different skin materials and experimental procedures used are given in the KSU study report [4].

EMPA Manikin

An adult-sized sweating thermal manikin, nicknamed SAM, has been developed at the Swiss Federal Laboratories for Material Testing and Research (EMPA) in Switzerland [5,6]. Sweat rates of up to 4 L/h, a maximum total heating power of over 1 kW, and realistic body movements characterize SAM's capabilities [7]. SAM has been used to assess the protection and comfort of clothing on a relative basis, e.g., [8]. Furthermore, results from SAM have been used to predict the dynamic physiological behavior of young male adults wearing different clothing [9]. Good agreement was observed with wear trial results for all clothing studied.

SAM has 30 separately-heated body segments and 125 sweating outlets distributed over its surface (Fig. 1*a*). For evaporative resistance measurements, a tightly-fitting skin is put onto SAM (Fig. 1*b*). This skin spreads water supplied to the outlets over the manikin surface by wicking. As hands and feet do not have sweat outlets, these parts are considered non-sweating but are heated to act as heat guards.

The present study compares results from this EMPA manikin with the interlaboratory results obtained previously and suggests possible ways to improve the reproducibility between labs.

As the manikins used by KSU and EMPA are both named SAM, the laboratory rather than the manikin abbreviations are used here to avoid confusion. Although KSU used two manikins for insulation measurements, only results from the SAM manikin of KSU are presented here.

Methods

Clothing Ensembles

For the previous interlaboratory study led by Kansas State University, five different clothing ensembles were measured at six different laboratories. Ensembles 2–5 consist of the base Ensemble 1 with additional clothing items added on top. The same ensembles, listed in Table 1, were measured using the EMPA manikin.

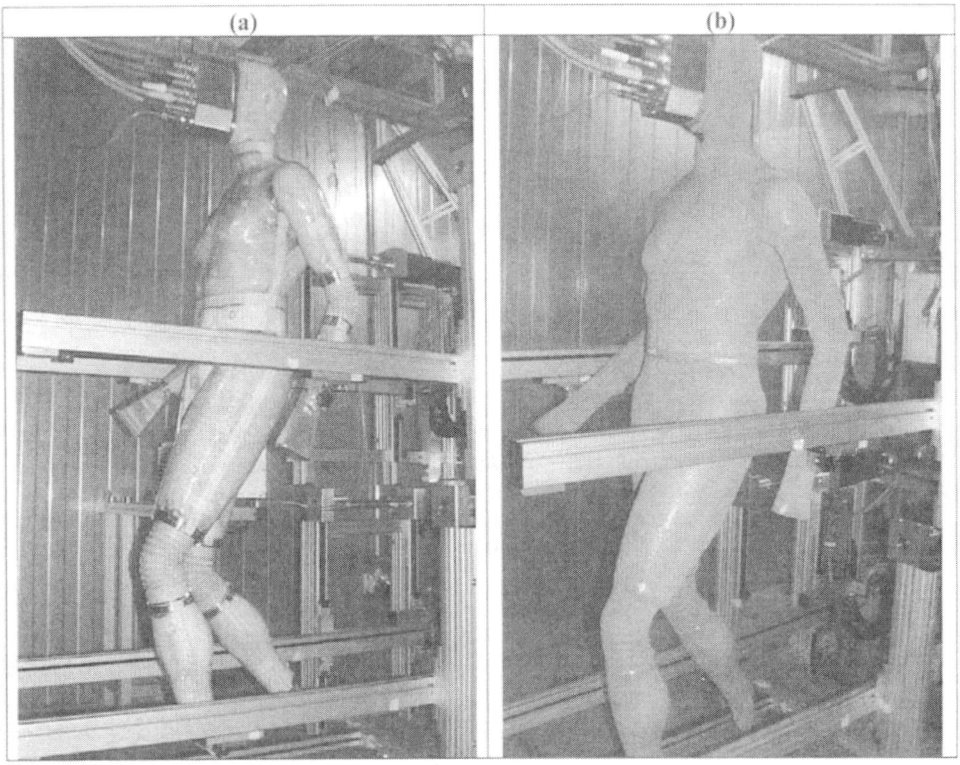

FIG. 1—*EMPA manikin SAM* (a) *without and* (b) *with the skin material. SAM is shown walking. For stationary results, the arms and legs hang unattached.*

TABLE 1—*Clothing ensembles measured.*

Ensemble	List of Garments
# 1. Warm weather (indoor)	Base ensemble: briefs, short-sleeve shirt, long trousers, socks
# 2. Cold weather (outdoor)	Base ensemble plus fiberfill jacket, mittens, hat
# 3. Flame resistant protective clothing	Base ensemble plus flame resistant long-sleeve shirt/jacket, long trousers, hood, FR knit gloves
# 4. Chemical protective clothing (level B suit)	Base ensemble plus chemical resistant long-sleeve shirt/jacket, long trousers, hood, thin gloves
# 5. Cold weather windproof, waterproof protective clothing	Base ensemble plus long underwear top and bottoms worn underneath, fleece top and bottoms, jacket and pants, fiberfill mittens, knit hat

Total Insulation, Total Thermal Resistance

The total thermal insulation is defined as the total resistance to dry heat transfer provided by the clothing and the "still" air layer surrounding the clothing. Dry heat is transferred by conduction, convection, and radiation. Following ASTM F 1291 "Measuring the Thermal

Insulation of Clothing with a Heated Manikin," the total thermal insulation in clo units, I_t, is defined as:

$$I_t = \frac{K \cdot (T_s - T_a) \cdot A}{H} \tag{1}$$

where:
I_t = total thermal insulation of the clothing plus the clothing surface air layer (clo);
K = conversion factor = 6.45 clo·W·m^{-2}·K^{-1};
T_s = mean surface temperature of the manikin (°C);
T_a = air temperature (°C);
A = heated surface area of the manikin (m^2); and
H = total power input to the heated manikin surface (W).

In the final draft standard ISO/FDIS 15831:2003(E) "Clothing – Physiological Effects – Measurement of Thermal Insulation by Means of a Thermal Manikin," two methods are given to determine the total thermal insulation, namely the parallel and serial methods. In this study, the parallel method is used to calculate all insulation values, where the heat losses, temperatures (area-weighted), and areas are summed [10].

The term "thermal resistance" has the same meaning as thermal insulation and is used interchangeably. Simplifying Eq 1 and defining the total thermal resistance in SI units gives:

$$R_t = \frac{(T_s - T_a) \cdot A}{H} \tag{2}$$

where R_t = total thermal resistance of the clothing plus the clothing surface air layer (m^2·K / W).

If the temperature sensors used are not directly on the surface of the manikin but are covered by a uniform insulating layer (such as a protective layer and possibly an additional tight-fitting textile skin, which does not form part of the clothing insulation to be measured), Eq 2 becomes:

$$R_t = \frac{(T_{ts} - T_a) \cdot A}{H} - R_s \tag{3}$$

where T_{ts} = mean average-weighted surface temperature measured by the temperature sensors (°C), and R_s = thermal resistance of the surface layer covering the temperature sensors (m^2·K / W).

For example, the manikins at VTT and NCSU measure T_{ts} using nickel resistance wire positioned over the skin but under netting for protection. Similarly, the EMPA manikin measures T_{ts} using nickel wire wound onto the plastic shell part surfaces. However this wire is covered by a thin vanish layer. Thus, the netting or the vanish act as an insulating layer, and R_s is inherently greater than zero. However, in the KSU study and in this study, R_s was assumed to be negligible (i.e., R_s = 0 and $T_{ts} = T_t$). Thus, Eq 2 was used to calculate R_t. Ideally, however, R_s should be quantified for each manikin to enable R_t to be determined more accurately.

Measurement Procedure for Insulation

For insulation measurements, EMPA followed the general procedures used by all labs except QinetiQ and BWU, with the air temperature set to 20°C for the nude manikin and Ensemble 1, 18°C for Ensembles 3 and 4, and 16°C for Ensembles 2 and 5. As the average surface temperature of each manikin was set to 34°C, the temperature gradients (T_{ts}–T_a) ranged from 14–18°C. Also, EMPA used the same relative humidity of 50 % as used by all labs apart from QinetiQ (11 % r.h.). All manikins apart from the QinetiQ manikin were located in climatic chambers with controlled temperature and relative humidity. The guidelines given for the tests were according to draft ISO/CD 15831 "Clothing – Physiological effects – Measurement of thermal insulation by means of a thermal manikin," except that the air velocity should be 0.3 m/s rather than 0.5m/s.

Intrinsic Clothing Insulation

The intrinsic (or basic) clothing insulation is related to the total thermal insulation by:

$$R_{cl} = R_t - \frac{R_a}{f_{cl}} = \frac{I_{cl}}{K} \tag{4}$$

where:
R_{cl} = intrinsic clothing insulation or resistance ($m^2 \cdot K$ / W);
R_a = insulation of the still air layer around the nude manikin ($m^2 \cdot K$ / W);
f_{cl} = clothing area factor; and
I_{cl} = intrinsic thermal insulation (clo).

The f_{cl} value can be estimated using ISO 9920 or measured using a photographic method [11]. For Ensembles 1–5, f_{cl} was estimated by comparing photographs of the ensembles to photographs used for the f_{cl} database at KSU. For an unclothed manikin with an insulating layer, the thermal resistance of the air layer is given as:

$$R_a = R_m - R_s \tag{5}$$

where R_m = the measured insulation of the nude manikin including the manikin surface layer and the surrounding air layer (m^2 K / W).

Evaporative Resistance

The draft ASTM standard "Standard Test Method for Measuring the Evaporative Resistance of Clothing Using a Sweating Manikin" (draft 2 submitted for ASTM F-23 subcommittee ballot in April 2003), which shall be referred to henceforth as the R_{et} draft standard, states two options to determine the total evaporative resistance. Here these options are redefined with an additional term to subtract the evaporative resistance of the manikin skin, R_{es} [12].
Thus option 1 becomes:

$$R_{et} = \frac{(P_s - P_a) \cdot A_s}{H_e} - R_{es} \tag{6}$$

where
R_{et} = total evaporative resistance of the clothing ensemble and surface air laver ($m^2 \cdot kPa$ / W);
P_s = water vapor pressure at the manikin's sweating surface (kPa);
P_a = water vapor pressure in the air flowing over the clothing (kPa);
A_s = area of the manikin's surface that is sweating (m^2);
H_e = heating power required to maintain the sweating areas at 34°C (W);
R_{es} = evaporative resistance of the manikin skin ($m^2 \cdot kPa$ / W);
and Option 2 becomes:

$$R_{et} = \frac{(P_s - P_a) \cdot A_s}{\lambda (dm/dt)} - R_{es} \tag{7}$$

where λ = heat of vaporization of water at the measured surface temperature (Wh/g), and dm/dt = evaporative rate of moisture leaving the manikin's sweating surface (g/h).

The evaporation rate (dm/dt) is equal to the sweat rate minus the rate of moisture buildup within the clothing. This rate of moisture buildup is determined by weighing the clothed manikin at regular time intervals during a test.

For the evaporative resistance values presented below, R_{es} was assumed to be negligible. Ideally however, as with R_s above, R_{es} should be quantified for each manikin, to enable R_{et} to be determined more accurately.

In accordance with ISO 11092 "Textiles – Physiological effects – Measurement of thermal and water-vapour resistance under steady-state conditions" and the R_{et} draft standard for manikin, the evaporative resistance R_{et} should be measured under isothermal conditions. Thus, as the average surface temperature of the manikin should be maintained at 34°C, an ambient air temperature of 34°C should also be used. However, in the KSU study, only BWU measured R_{et} under isothermal conditions. KSU measured the nude manikin with its skin and the ensembles with a temperature gradient of 6°C. The other four labs used non-isothermal conditions with the same ambient temperatures for each ensemble as used for the dry insulation tests given above. EMPA used the same non-isothermal conditions as these four labs. Apart from BWU, which used Option 1, all other labs including EMPA used Option 2 to calculate R_{et}.

All manikins had a wetted textile layer covering most or the entire manikin surface to spread water laterally over the surface by wicking. In this way, the evaporation of sweat from the human skin was simulated. However, the textile material used, the technique of wetting, and the measurement procedure adopted to determine R_{et} varied considerably from lab to lab.

For the EMPA manikin, a constant sweat rate of 300 g h^{-1} (272 g $h^{-1}m^{-2}$) over a 3-h period was used to determine the evaporative resistance of each ensemble. The total weight of the manikin was measured every minute throughout this period. The sweat and condensation rates were calculated using 30-min intervals, and the last 10 values for each rate were averaged. The relative humidity close to the skin was measured at the chest and the lower back using omni-directional Hygroclip SC05 sensors from Rotronic AG, Switzerland with an accuracy of ±1.5 % rh. This enabled estimation of the sweat rate and time of sweating required to reach saturation at the skin (indicated by a relative humidity of 100 %).

Analogous to the intrinsic thermal insulation given in Eq 4, the intrinsic evaporative clothing resistance is defined as:

$$R_{ecl} = R_{et} - \frac{R_{ea}}{f_{cl}} \tag{8}$$

where:
R_{ecl} = intrinsic evaporative clothing resistance (m²·kPa / W);
R_{ea} = evaporative resistance of the air layer around the nude manikin (m²·kPa / W); and
f_{cl} = clothing area factor.

Analogous to Eq 5, the evaporative resistance of the air layer R_{ea} is given as:

$$R_{ea} = R_{em} - R_{es} \tag{9}$$

where R_{em} = evaporative resistance of the manikin including the skin and the air layer (m²·kPa / W). The R_{em} value is measured for each manikin with only the wetted tight-fitting skin on.

Further details of the manikins and methods used by labs involved in the KSU study to determine R_t and R_{et} have been given previously [13–16].

The moisture permeability index indicates the maximum evaporative heat transfer permitted by a clothing system compared to an ideal maximum from an uncovered surface. This is defined as [17]:

$$i_m = \frac{R_t / R_{et}}{R_a / R_{ea}} = \frac{R_t / R_{et}}{16.65} \qquad (11)$$

where i_m = permeability index for the clothing including the air layer resistance, and R_a / R_{ea} is taken to have the typical value of 16.65°C/kPa (Lewis relation).

Similarly, the moisture permeability for the clothing alone is defined as follows:

$$i_m = \frac{R_{cl} / R_{ecl}}{16.65} \qquad (12)$$

Results

The thermal resistance of the nude manikin (R_m) measured for each nude manikin in the KSU study varied from 0.051–0.104 m^2·K/W compared to an EMPA value of 0.062 m^2·K/W. The evaporative resistance of the manikin with skin (R_{em}) varied from 0.011–0.032 m^2·kPa/W in the KSU study compared to an EMPA value of 0.022 m^2·kPa/W. Thus, the manikin resistances of SAM at EMPA lie within the ranges of those for the other manikins. The spread of results is attributed to the differences in the resistances of the manikin surfaces (R_s and R_{es}) and differences in the resistances of the air layer (R_a and R_{ea}). Differences in the resistances of the air layer are primarily due to the different wind speeds used at the different labs. Since the manikin resistances varied so much from lab to lab and the manikin surface resistances were assumed to be negligibly small in this study, the intrinsic rather than the total resistance values are studied here.

For each ensemble, three separate measurements were carried out at each lab. Using the above equations, average values of the intrinsic thermal and evaporative resistance values and permeabilities were calculated (assuming $R_s = 0$ and $R_{es} = 0$). These values are listed in Table 2.

The R_{cl} and R_{ecl} values tend to increase with the number of clothing layers with Ensemble #1 having the lowest values and Ensembles #4 and #5 the highest. Comparing results between the different labs for each ensemble showed that QinetiQ consistently obtained the lowest values for R_{cl} and R_{ecl} and NCTRF consistently the highest R_{cl} values. Other systematic rankings of manikin results are not evident. It was found easier to obtain better agreement between labs with ensembles with lower resistance values (Ensembles #1 to #3) than with higher resistance values (Ensembles #4 and #5).

Generally the EMPA results compare favorably with the results from other labs. For example, the EMPA results for R_{cl} and R_{ecl} lie well within the range of values obtained elsewhere. The standard deviations (not listed) are mainly smaller than for the other laboratories, with values ranging from 0.1–0.5 % for R_{cl} and from 2–5 % for R_{ecl} of all ensembles apart from Ensemble 4, which gave a much higher standard deviation of 42 %. Ideally, the evaporative measurements for Ensemble 4 should have been repeated, but time did not allow for this.

In the interlaboratory KSU study, statistics were calculated according to ASTM E 691, "Standard Practice for Conducting an Interlaboratory Study to Determine the Precision of a Test Method." Resistance values of all the labs involved were used to calculate the 95 % repeatability limit, **r** (an indicator of the variability between independent test results obtained on a specific set of samples within a single laboratory) and the 95 % reproducibility limit, **R** (an indicator of the variability between single test results obtained at different laboratories). For example, an **R** of 0.100 m^2·K/W would mean than 95 % of the time, the thermal resistance values measured for a particular type of ensemble at different labs would not differ more than ±0.100 m^2·K/W.

TABLE 2—*Intrinsic resistance values.*

Ensemble (Clothing Area Factor)	Laboratory	Intrinsic Dry Thermal Resistance		Intrinsic Evaporative Resistance	Permeability Index
		R_{cl} ($m^2 \cdot K/W$)	I_{cl} (clo)	R_{ecl} ($m^2 \cdot kPa/W$)	i_{cl}
#1	QinetiQ	0.064	0.41	0.010	0.39
Warm Indoor Clothing	NCTRF	0.151	0.97	0.016	0.57
(f_{cl} = 1.30)	VTT	0.113	0.73	0.016	0.44
	BWU	0.099	0.64	0.020	0.30
	KSU	0.104	0.67	0.019	0.33
	NCSU	0.122	0.79	0.011	0.66
	EMPA	0.107	0.69	0.011	0.58
#2	QinetiQ	0.186	1.20	0.030	0.37
Cold Outdoor Clothing	NCTRF	0.288	1.86	0.035	0.51
(f_{cl} = 1.40)	VTT	0.230	1.48	0.043	0.33
	BWU	0.203	1.31	0.038	0.32
	KSU	0.274	1.77	0.044	0.38
	NCSU	0.248	1.60	0.033	0.46
	EMPA	0.267	1.72	0.028	0.57
#3	QinetiQ	0.113	0.73	0.016	0.44
Flame Resistant	NCTRF	0.211	1.36	0.022	0.59
Protective Clothing	VTT	0.174	1.12	0.026	0.41
(f_{cl} = 1.35)	BWU	0.155	1.00	0.028	0.34
	KSU	0.181	1.17	0.028	0.39
	NCSU	0.195	1.26	0.022	0.54
	EMPA	0.190	1.23	0.020	0.57
#4	QinetiQ	0.156	1.00	0.083	0.12
Chemical Protective	NCTRF	0.251	1.62	0.289	0.05
Clothing	VTT	0.199	1.28	0.188	0.07
(f_{cl} = 1.50)	BWU	0.215	1.39	0.245	0.05
	KSU	0.225	1.45	0.090	0.15
	NCSU	0.214	1.38	0.201	0.07
	EMPA	0.223	1.44	0.152	0.09
#5 Cold weather	QinetiQ	0.294	1.90	0.085	0.21
windproof and	NCTRF	0.425	2.74	0.087	0.30
waterproof clothing	VTT	0.318	2.05	0.181	0.11
(f_{cl} = 1.45)	BWU	0.277	1.79	0.154	0.11
	KSU	0.344	2.22	0.091	0.23
	NCSU	0.323	2.08	0.296	0.07
	EMPA	0.374	2.41	0.150	0.15

Precision statistics were recalculated for all the results including the EMPA results. QinetiQ consistently obtained the lowest insulation and evaporative resistance values due to the very different measurement conditions used, e.g., a much higher air flow of 1.55 m/s for R_t as opposed to 0.2–0.6 m/s used elsewhere and a much lower relative humidity of 11 % instead of 50 %. Consequently, precision statistics including EMPA but excluding QinetiQ results were also calculated.

Mean values and precision statistics for R_{cl} are compared in Table 3. Both in-lab repeatability and inter-lab reproducibility tend to vary with the magnitude of the R_{cl} value. For all ensembles, inclusion of EMPA data and exclusion of QinetiQ data cause the mean values to increase progressively. Including EMPA results causes the repeatability limits to reduce slightly for all ensembles, except for Ensemble 1, where no change is seen. The reproducibility improves with inclusion of EMPA results and improves even more with exclusion of QinetiQ results.

In Table 4, mean values and precision statistics for R_{ecl} are compared. Similar to the R_{cl} precision statistics, the in-lab repeatability and the inter-lab reproducibility vary with the magnitude of the R_{ecl} value. Except for Ensemble 3, the repeatability values improved with EMPA results included. Also, the reproducibility improved with inclusion of EMPA results and even more with exclusion of QinetiQ results. Average relative humidity values measured next to the skin at the chest are shown in Fig. 2. The average values measured for the back reached 100 % r. h. faster than for the chest.

TABLE 3—*Repeatability (**r**) and Reproducibility (**R**) for Mean Intrinsic Thermal Resistance R_{cl} ($m^2 \cdot K/W$).*

Ensemble	Study Excluding EMPA			Study of All Laboratories			Study Excluding QinetiQ		
	Mean	r	R	Mean	r	R	Mean	r	R
# 1	0.109	0.009	0.081	0.109	0.008	0.074	0.116	0.008	0.053
# 2	0.238	0.025	0.113	0.242	0.023	0.108	0.251	0.022	0.090
# 3	0.172	0.019	0.097	0.174	0.018	0.090	0.184	0.018	0.056
# 4	0.210	0.014	0.084	0.212	0.013	0.082	0.221	0.014	0.050
# 5	0.330	0.053	0.152	0.336	0.049	0.146	0.343	0.053	0.150

TABLE 4—*Repeatability (**r**) and Reproducibility (**R**) for Mean Intrinsic Evaporative Resistance R_{ecl} ($m^2 \cdot kPa/W$).*

Ensemble	Study Excluding EMPA			Study of All Laboratories			Study Excluding QinetiQ		
	Mean	r	R	Mean	r	R	Mean	r	R
# 1	0.015	0.003	0.012	0.015	0.002	0.012	0.015	0.002	0.012
# 2	0.037	0.006	0.016	0.036	0.005	0.018	0.037	0.004	0.018
# 3	0.023	0.005	0.014	0.023	0.005	0.013	0.024	0.005	0.010
# 4	0.183	0.081	0.241	0.178	0.080	0.224	0.194	0.086	0.207
# 5	0.149	0.058	0.235	0.149	0.055	0.215	0.160	0.059	0.219

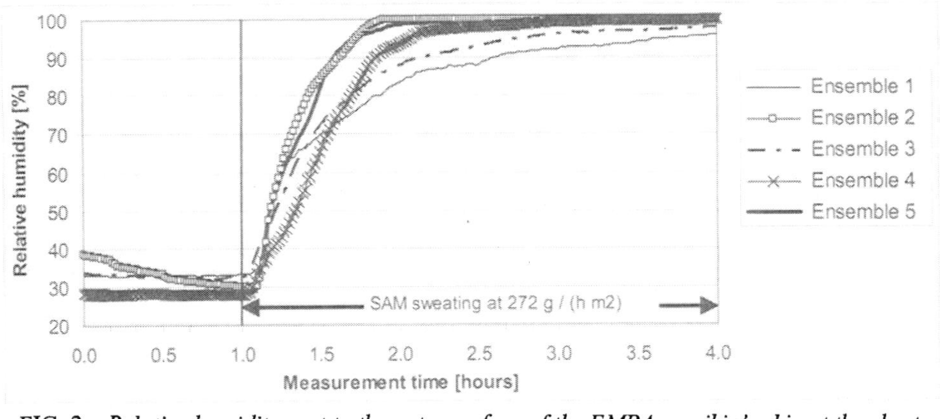

FIG. 2—*Relative humidity next to the outer surface of the EMPA manikin's skin at the chest.*

The moisture distribution in the tight-fitting suit is not completely uniform due to water supplied from each sweat outlet not wicking out completely to cover all areas of the manikin to the same extent. However, these measurements indicate that a very high humidity has been reached close to the skin for all suits. The relative humidity was assumed to have reached 100 % (skin saturated) after 2½ h of sweating had elapsed for all ensembles.

Discussion

Thermal Resistance

Results for R_{cl} presented above were obtained without the tight-fitting skin. Additional measurements where made with the skin left on underneath the clothing. The intrinsic thermal insulation of the skin (21.7±0.4 $m^2 \cdot K/W$) was subtracted from the total thermal insulation values obtained, to determine the approximate intrinsic thermal insulation value of each ensemble. Intrinsic values obtained in this way were 0.2–8.3 % lower than those measured without the skin.

Furthermore, the standard deviations were higher when the skin was present ranging from 0.2–2.1 %. These differences can be attributed to the uncertainty of the exact value of the intrinsic thermal insulation of the skin and the different surface roughness of the skin compared to the outer surface of each ensemble, which alters the effective thermal insulation of the air layer R_a. Thus measuring the thermal insulation of the clothing without the tight-fitting skin is the preferred method to obtain accurate results.

The above differences in results obtained with and without the skin demonstrate the effect of neglecting the thermal resistance of the manikin surface, R_s, defined in Eq 3. With $R_s = 0$, the R_t and R_a values are too large, and consequently the R_{cl} value will then be calculated as too low from Eq 4. When a skin is used, the temperature sensors are covered by this insulating layer and $R_s > 0$. Thus, particularly for a sweating manikin, which cannot have its textile skin removed for practical reasons, R_s must be determined and used to calculate R_t. However, even for a nude manikin with its skin removed, R_s may well be greater than zero.

Evaporative Resistance

For the evaporative resistance measurements, additional measurements were made at EMPA with a lower wind speed of 0.3 m/s and a shorter sweating duration of 2 h. However, the R_{ecl} values were from 0.5–2 times greater. These large differences are mainly due to the fact that steady state was not reached during these 2 h. Also, a lower sweat rate of 200 g h^{-1} m^{-2} was used for these additional measurements. For Ensemble 1, the average relative humidity measured close to the skin was less than 90 % and starting to reach steady state at the end of the 2 h. Thus, the assumption that the skin surface was saturated was not valid in this case. These observations demonstrate the importance to ensure that: (a) steady state has been reached, and (b) that the sweat rate is sufficiently high to saturate the skin. The optimal sweat rate may be different for different clothing as suggested in Fig. 2. However, a systematic study to determine such optimal sweat rates for different types of clothing has not yet been carried out. Nevertheless the use of humidity probes close to the skin seems a useful way of estimating when saturation of the skin occurs for a particular ensemble without having to remove the clothing for visual inspection.

As for R_s discussed above, the evaporative resistance of the skin R_{es} may not be negligible and must be determined.

As with the manikins from VTT and NCSU, the EMPA manikin was not made to sweat at the head, hands, and feet. The other manikins were wetted over the whole surface. This will have affected the evaporative resistance values obtained for these manikins to some degree. Ideally, as specified in the R_{et} draft standard, the whole of the manikin surface should be wetted.

At KSU, a thin cotton skin covering the manikin was sprayed with water and dressed quickly. As this skin dried slowly with time, steady-state conditions were not reached over the 45-min measurement period, and R_{et} was estimated from measurements at near to steady-state conditions. The other manikins used various mechanisms to continually supply water to the surface layer.

Such a constant sweat rate throughout the measurement is required for steady-state conditions and accurate determination of the evaporative resistance.

Sweat Water Temperature

For the tests in this study, the water used to simulate sweat on the manikin's surface was not specified to have a particular temperature, which may have affected the reproducibility somewhat. In the draft R_{et} standard, the water temperature is standardized to 35 ±0.5°C.

Airflow

In the climate chamber used at EMPA for these tests, reducing the airflow from 0.6 m/s down to 0.3 m/s caused fluctuations in relative humidity over time to increase from about ±1 % to up to ±3 %. Similar time-related fluctuations in the heating power at steady state increased typically from ±1 % to ±1.5 % of the given mean value. Some optimization of the climate control should be possible to reduce these fluctuations and possibly improve uniformity for the lower airflow of 0.3 m/s. In fact, the R_{et} draft standard proposes using a maximum air speed of 0.15 m/s, for which climate stability and uniformity may prove more difficult. However, the dry thermal insulation standards define higher air speeds (e.g., 0.5±0.1 m/s for ISO/FDIS 15831 and 0.4±0.1 m/s for ENV 342, "Protective clothing against cold").

Conclusions

Including the EMPA results with the interlaboratory data collected previously improved the reproducibility of the intrinsic thermal resistances by 4–9 % and of the intrinsic evaporative resistances by 7–9 % for all ensembles measured. Including the EMPA results but excluding the QinetiQ results improved the reproducibility of the intrinsic thermal resistance by 14–61 % and of the intrinsic evaporative resistance for three ensembles by 7–29 % compared to the KSU study. Therefore, the EMPA results benefit the overall interlaboratory reproducibility. Furthermore, the QinetiQ results should be removed from this database of results, as the climatic conditions and methodologies used were very different than for all the other manikins.

In spite of these reductions of the reproducibility values, finite values are still very large ranging from 22–46 % for the intrinsic thermal resistances and from 41–138 % for the intrinsic evaporative resistances. These large variabilities between labs are due to the differences in manikin design and the measurement and calculation methodologies used.

Possible improvements in the methodology used to determine the thermal resistance include:

- Removal of the textile skin layer for dry tests when possible.

- Quantification of the thermal resistance of the manikin surface, R_s.

Possible improvements in the methodology used to determine the evaporative resistance include:

- Quantification of the evaporative resistance of the manikin surface, R_{es}.
- Possible use of humidity sensors next to the surface of the wetted skin to indicate saturation of the skin.
- Sweating over the whole manikin surface is desirable.
- Constant sweating is required for steady-state conditions.

Another interlaboratory round robin test is planned following the development of the R_{et} draft standard. Particular emphasis will be placed on using the same experimental procedure in all labs, including the proposed use of isothermal conditions. In this way it is hoped that both the repeatability within each lab and the reproducibility between labs can be improved.

References

[1] Holmer, I., "Manikin History and Applications," *The 5th International Meeting on Manikins and Modelling, 5I3M*, V. Candas, Ed., Centre d'Etudes de Physiologie Appliquée (CEPA), Strasbourg, France, 2003, CD-ROM.

[2] Elabbassi, E. B., Belghazi, K., Delanaud, S., and Libert, J., "Dry Heat Loss in Incubator: Comparison of Two Premature Newborn Sized Manikins," *The 5th International Meeting on Manikins and Modelling, 5I3M*, V. Candas, Ed., Centre d'Etudes de Physiologie Appliquée (CEPA), Strasbourg, France, 2003, CD-ROM.

[3] Kuklane, K., Holmér, I., Tochihara, Y., Fukazawa, T., Lee, G., Nonaka, T., et al., "Comparison of Heat Transfer from Baby and Adult Manikins," *The 5th International Meeting on Manikins and Modelling, 5I3M*, V. Candas, Ed., Centre d'Etudes de Physiologie Appliquée (CEPA), Strasbourg, France, 2003, CD-ROM.

[4] McCullough, E. A., "An Interlaboratory Study of Manikins Used to Measure the Thermal and Evaporative Resistance of Clothing," Institute for Environmental Research (IER) Technical Report #02-04, Kansas State University, MA, January 2002.

[5] Richards, M. G. M. and Mattle, N. G., "Development of a Sweating Agile Thermal Manikin (SAM)," *4th International Meeting on Thermal Manikins, 4IMM*, M. G. M. Richards, Ed., Swiss Federal Laboratories for Materials Testing and Research (EMPA), St. Gallen, Switzerland, 2001, CD-ROM.

[6] Zimmerli, T., "Manikin Testing of Protective Clothing - A Survey," *Performance of Protective Clothing: Issues and Priorities for the 21st Century: Seventh Volume, ASTM STP 1386*, C. N. Nelson and N. W. Henry, Eds., ASTM International, West Conshohocken, PA, 2000.

[7] Richards, M. G. M. and Mattle, N. G., "A Sweating Agile Thermal Manikin (SAM) Developed to Test Complete Clothing Systems under Normal and Extreme Conditions," *Human Factors and Medicine Panel Symposium - Blowing Hot and Cold: Protecting Against Climatic Extremes*, RTO/NATO, ISBN 92-837-1082-7, Vol. 4, Dresden, 8–11 October 2001, pp. 1–7.

[8] Richards, M. G. M., Mattle, N. G., and Becker, C., "Assessment of the Protection and Comfort of Fire Fighters' Clothing Using a Sweating Manikin," *2nd Eurpoean Conference on Protective Clothing (ECPC) and NOKOBETEF 7*, H. Mäkinen, R. Rossi, Eds., Montreaux, Switzerland, 2003.

[9] Richards, M. G. M., "Modelling Fire-Fighter Responses to Exercise and Asymmetric IR-Radiation Using a Dynamic Multi-Mode Model of Human Physiology and Results from the Sweating Agile Thermal Manikin (SAM)," *The 5th International Meeting on Manikins and Modelling, 5I3M*, V. Candas, Ed., CEPA, Strasbourg, France, 2003, CD-ROM.

[10] Nilsson, H., "Analysis of Two Methods of Calculating the Total Insulation," *Proceedings of a European Seminar on Thermal Manikin Testing at the National Institute for Working Life*, Arbete och Hälsa, 1997, Vol. 9, pp. 17–22.

[11] McCullough, E. A., Jones, B. W., Byron, W., and Huck, J., "A Comprehensive Data Base for Estimating Clothing Insulation," *ASHRAE Transactions*, Vol. 91, No. 2, 1985, pp. 29–45.

[12] Fan, J. and Chen, Y. S., "Measurement of Clothing Thermal Insulation and Moisture Vapour Resistance Using a Novel Perspiring Fabric Thermal Manikin," *Measurement Science and Technology*, Vol. 13, 2002, pp. 1115–1123.

[13] Higenbottam, C., "Modification of a Thermal Manikin for Measurement of Water Vapour Permeability of Clothing," *The 7th International Conference on Environmental Ergonomics*, Jerusalem, Israel, 1996, pp. 409–412.

[14] McCullough, E. A., Jones, B. W., and Tamura, T., "A Data Base for Determining the Evaporative Resistance of Clothing," *ASHRAE Transactions*, Vol. 91, 1989, pp. 316–328.

[15] Meinander, H., "Extraction of Data from Sweating Manikin Tests," *Proceedings of the Third International Meeting on Thermal Manikin Testing, 3IMM*, Arbete och Hälsa, Vol. 4, 2000, pp. 95–99.

[16] Tamura, T. and Tomizawa, M., "Thermal Control System for a Sweating Manikin," *Journal of Home Economics in Japan*, Vol. 44, No. 8, 1993, pp. 5671–5677.

[17] Woodcock, A. H., "Moisture Transfer in Textile Systems, Part I," *Textile Research Journal*, Vol. 32, 1962, pp. 628–633.

Journal of ASTM International, February 2005, Vol. 2, No. 2
Paper ID JAI12105
Available online at www.astm.org

Donna H. Branson,[1] *Cheryl A. Farr,*[1] *Semra Peksoz,*[1] *Jinhee Nam,*[1] *and Huantian Cao*[1]

Development of a Prototype Personal Cooling System for First Responders: User Feedback

ABSTRACT: A series of six focus groups in five U.S. cities was conducted as a component in a 3-year project to develop a prototype personal cooling system for first responders. The target user was a HAZMAT worker wearing level A and B ensembles, with a 30-min mission and an average body heat dissipation of 180 W. The purpose of the focus groups was to determine user needs, wants, and concerns early and progressively throughout the development of the prototype system to serve as design input. The first two focus groups provided information on the use of HAZMAT ensembles and thermal comfort problems associated with wearing the ensembles while working an incident, and reaction to a mock-up cooler system. Focus groups 3–6 provided user feedback on an alpha prototype system. All groups confirmed the need for a cooling system and the problem with moisture build-up within level A and B encapsulated suits. Reaction to the liquid-cooled vest was positive. An exterior location for the cooler unit was acceptable if users' concerns were addressed adequately.

KEYWORDS: cooling system, liquid-cooling garment, level A chemical protection, level B chemical protection, user feedback, thermal comfort

Introduction

The first responder community, that is, local police, firefighters, and emergency medical professionals, represent America's first line of defense in a terrorist attack, a natural disaster, or any emergency situation. There are over one million firefighters in the United States, and more than half are volunteers. Local police departments and sheriffs' offices together represent almost 900 000 full-time employees, and there are over 155 000 nationally registered emergency medical technicians (EMT) in the United States. "Properly trained and equipped first responders have the greatest potential to save lives and limit casualties" [1].

First responders' missions vary considerably, including cleanup of a hazardous materials spill at an industrial site or train derailment; worker decontamination; seizure and cleanup of a methamphetamine lab; structural fires; and search and rescue missions following a terrorist attack, natural disaster, or other emergency. The nature of the hazard dictates the type of personal protective equipment (PPE) needed by the first responder [2,3]. Thus today, a wide choice of PPE is available to the first responder community; each is designed to provide a level of protection against hazards such as fire, heat, chemicals, biological agents, etc. In incidents involving unknown hazardous substances, it is common practice to use PPE that provides the highest level of protection. For liquid and vapor chemical protection, this means a level A suit (a fully encapsulated garment), a self-contained breathing apparatus, and ancillary equipment.

Manuscript received 18 November 2003; accepted for publication 9 July 2004; published February 2005. Presented at ASTM Symposium on Performance of Protective Clothing: Global Needs and Emerging Markets: 8th Symposium on 13 January 2004 in Tampa, FL.

[1]Professor, Associate Professor, Graduate Student, Graduate Student, and Assistant Professor, respectively, Oklahoma State University, Department of Design, Housing and Merchandising.

Table 1 contains a brief review of the chemical protection levels A and B as suggested by the Environmental Protection Agency.

TABLE 1—*PPE for chemical protection levels A and B [4].*

Description	Protection Provided	Used When	Limitations
LEVEL A: Vapor protective suit (meets NFPA 1991). Pressure-demand, full-face SCBA, inner chemical-resistant gloves and safety boots, 2-way radio communication; optional: cooling system, outer gloves, hard hat.	Highest available level of respiratory, skin, and eye protection from solid, liquid, and gaseous chemicals.	Chemicals are identified and have a high level of hazard to respiratory system, skin, and eyes; substances are present with known or suspected skin toxicity or carcinogenity; operations conducted in confined and/or poorly ventilated areas.	Protective clothing must resist permeation by the challenge chemical or mixture. Ensemble items must allow integration without loss of performance.
Level B: Liquid splash-protective suit (meets NFPA 1992); pressure-demand, full-facepiece SCBA; inner chemical-resistant gloves and safety boots, 2-way radio communication, hard hat; optional: outer gloves, cooling system.	Same respiratory as A, but less skin protection; liquid splash protection, but no protection against chemical vapors or gases.	Chemicals are identified but do not require a high level of skin protection. Initial site surveys are required until higher levels of hazards are identified. Primary hazards associated with site entry are from liquid and not vapor contact.	Protective clothing items must resist penetration by the challenge chemicals or mixtures; ensemble items must allow integration without loss of performance.

Purpose

This study was undertaken as a component in a 3-year project to develop a personal cooling system for first responders wearing level A and B protective ensembles. Figure 1 presents a flow chart of the major project components and their relationships for this phase of the project. The two-fold purpose of the study reported in this paper was to understand activities and needs of the HAZMAT responder, especially in relation to heat stress, and to obtain their input as part of the product development process. To that end, a series of six focus groups was held over a 15-month period in five locations. The first two focus groups were completed early in the process as a part of the *garment research* component, as shown in Fig. 1, with the resulting data used in the development of the alpha garment prototype. The last four focus groups, which were held after the alpha prototype cooler and garment were fabricated and their effectiveness was determined, provided user feedback on the prototype system. This feedback was used in the refinement of the cooler system for subsequent testing of the system using human subjects.

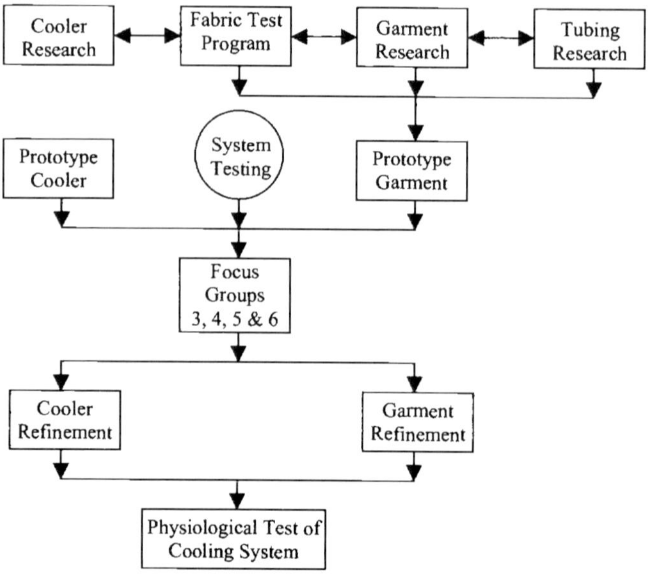

FIG. 1—*Research components resulting in the design, fabrication, testing and user feedback of an alpha prototype personal cooling system.*

Method

Focus Group as a Data Collection Method

By definition, a focus group is a group of individuals who can provide insights or in-depth knowledge about a particular object or event. Therefore, the focus group was selected as the method for obtaining user input for the product development efforts. The strengths of this method include: 1) having a pre-planned question protocol for each focus group, thereby providing an opportunity to collect data about specific points or areas of interest; 2) having the ability to ask follow-up probe questions, thus providing an opportunity to explore unanticipated feedback or to explore an idea in greater depth; 3) having potential users provide spontaneous insights into the strengths, limitations, and anticipated problems with the product under development, thus identifying alternative uses for the product, unacceptable aspects of the product, and unanticipated problems; 4) having sequential focus groups as the product development process progresses, thereby providing ongoing information as issues and problems identified early are solved and new problems emerge; and 5) having the opportunity for input from multiple groups, thereby receiving confirmation of findings emerging in the process and identifying concerns unique to specific groups.

The specific protocol used for this research began with all team participants submitting questions of interest that were compiled into a questionnaire in which the questions were organized by topic and reviewed by the research team, and modifications were made. The Bureau for Social Research at Oklahoma State University moderated each focus group, recorded the discussion on audiotapes, and prepared transcriptions of the tapes. The primary function of the moderator was to ensure that all questions on the questionnaire were included in the discussion and that all participants had an opportunity to respond. At least two research team members were present at all focus groups to assist the moderator with probe questions. The transcripts were used by the research team trained in coding for content analysis to identify relevant information in the responses. The facts identified in the content analysis were organized by topic or into themes. At least two members of the research team analyzed each transcript, thereby minimizing the potential for individual bias. Comparisons were made between the coders to identify individual variation on the coding of the first focus group. The interrater reliability was calculated using the formula [5]:

$$\frac{\text{Number of agreements}}{\text{Number of agreements} + \text{Number of disagreements}} \times 100 \qquad (1)$$

The interrater reliability was 92.3 %. The Oklahoma State University Institutional Review Board approved the protocol.

Instruments

In total, three instruments were developed with questions as well as probe follow-up questions as given in Table 2. The first instrument focused on questions related to types of HAZMAT gear worn, frequency of use, the environments in which the suits were worn, the types of incidents for which each type of suit was used, the first responders' heat stress problems, and other associated problems related to incorporating a cooling garment into the HAZMAT ensemble. The second instrument focused on questions related to identifying user needs, wants, and concerns related to personal cooling devices, particularly in regard to the acceptability of a mock-up cooling unit, possible locations for the cooler unit and potential attachment methods, the interface between the cooler unit and the cooling garment, and the design of the cooling garment. The third instrument was designed to obtain feedback on the research team's alpha prototype vest and cooler system. This instrument was used for focus groups in four selected cities.

Sample

For the first two focus groups, subjects were recruited from fire departments in three Midwestern cities, specifically Oklahoma City and Stillwater, Oklahoma and Wichita, Kansas, and from a Midwestern petroleum company. Eight male subjects participated in focus group 1, and six male subjects participated in focus group 2. Four of the eight subjects who participated in the first focus group also participated in the second focus group. All volunteers were recruited by contacting an appropriate senior level administrator and requesting participation by those trained in HAZMAT operations. Subjects in focus group 1 ranged in age from 38–46 with a mean age of 43 years. Of the eight subjects in focus group 1, five had completed college, two had an associate's degree, and one had college coursework. As a group, they had 6–21 years of HAZMAT experience, including holding various leadership positions. Subjects in focus group 2 ranged in age from 38–47 with a mean age of 43 years. As a group, they had considerable HAZMAT experience and training, including leadership positions with the National Fire Protection Association (NFPA).

TABLE 2—*Focus group questions.*

Focus Group 1	Focus Group 2	Focus Groups 3–6
Explain that the purpose of today's meeting is to ascertain the use of the HAZMAT suits, the environments in which the suits are used, and perceptions related to cooling devices.	Explain that the purpose of today's meeting is to get feedback concerning the cooling unit and characteristics of the various components.	Explain that the purposes of today's meeting is to ascertain 1) the use of the HAZMAT suits, the environments in which the suits are used, and perceptions related to cooling devices; and (2) to get feedback concerning the cooling unit and characteristics of the various components.
Approximately how many times in a year did your unit don HAZMAT Level A suits in response to an incident? Level B?	Request feedback related to the size and weight of the cooler.	a. Approximately how many times in a year did your unit don HAZMAT Level A suits in response to an incident? Level B?
Please explain the "incidents" that generated response from the HAZMAT team over the past year?	Request feedback related to possible location for the pass-through to connect the vest manifold to the cooler.	How important is personal cooling to you for conducting your work?
At what points in the HAZMAT process would it be "nice" to have cooling? When is it essential?	Request feedback related to the preferred location of the cooling unit.	What is your experience with cooling vests? What are the advantages and limitations?
What is the field of vision from within a HAZMAT suit? Are blind controls on a cooling device necessary?	Request feedback related to how the cooling unit could be attached to the HAZMAT suit.	Show prototype cooler: a. Request feedback related to the size and weight of the cooler. b. What are your concerns relative to the cooler unit being located outside the suit? c. What system weight is acceptable?
During the past year, describe the different decontamination procedures that your team used.	Request feedback on the liquid cooling vest regarding the location of the tubing relative the SCBA.	Show prototype garment: a. What are your thoughts about the placement of the manifold at the top (neck) or bottom of the vest? b. What are your thoughts about the closure system in the garment front? Garment shoulder?
What is your experience with cooling vests? What are the advantages and limitations? Explain the respirator system that you use for Level A and B suits.	Request information to determine if a T-shirt is routinely worn under the SCBA. Request information to determine if the subjects would wear a cooling vest next to the skin. Request information to determine if the subjects would wear a "snug" fitting cooling vest.	Do you think the system meets your requirements for ease of donning and doffing?

For focus groups 3–6, an individual operations unit was contacted in each selected location with the assistance of the project technical manager and Oklahoma State University alumni contacts. The cities selected for the last four focus groups provided diversity by geography, climate, and location. Las Vegas and Phoenix are representative of units that operate in low humidity environments with temperatures that are frequently in excess of 37.8°C (100°F). Houston is representative of a high humidity environment with temperatures that are frequently over 32.2°C (90°F). New York City was selected due to recent first responder experiences, and because temperatures are less intense there than in other locations. Table 3 presents the number of participants in each of the four cities by gender, age range, and mean age. All had served with the Fire Service and Special Operations for a considerable number of years, and as such, they had considerable HAZMAT training.

TABLE 3—*Focus group demographics.*

City	N	Gender	Age Range	Mean Age
Las Vegas	10	9 M, 1 F	34–56	45.4
Phoenix	8	8 M	34–54	44.5
Houston	10	10 M	30–50	40.5
New York	8	8 M	29–47	39.8

Results and Discussion

Focus Group 1

Figure 2 contains focus group 1 participants' responses to questions regarding frequency of use of level A and B suits and the incidents that required the use of the suits.

Level A suits used 2–3 times per year for incidents, typically for 1 h or less duration.
Level A suits worn for training more than for responding to incidents.
Level B suits used more than 100 times per year for incidents, typically for longer time periods.
Level A suits worn daily by some industry and/or lab workers dealing with highly toxic materials or unknown substances.
Level B suits worn by decontamination teams and methamphetamine lab seizure teams.

FIG. 2—*Focus group 1: frequency of use of level A and B suits and incidents.*

Thus, level B suits were used by the participants to a greater extent than level A suits, and they were worn for a longer time period. Two reasons were given for the reduced frequency of level A usage: 1) greater care and more education in the work force is resulting in a reduced number of incidents; and 2) HAZMAT teams are more educated, allowing them to recognize and assess an incident with more precision and to select their suits accordingly. Level A is typically used when a substance is unknown. It should be noted that length of wearing level A suits is also influenced by the air supply, which typically consists of 30- or 60-min air bottles, except for industry teams that frequently have an airline. Respondents noted that a 60-min bottle would typically last 30–40 min depending on the individual and the work he or she would perform. A summary of the problems reported for level A and B suits is shown in Fig. 3.

Neither level A nor B suits provide fire/flash protection.
Level A suit and respirator are heavy, inflexible, and stiff.
Level A suits often inflate, further restricting movement.
Humidity build-up inside encapsulated suits produces discomfort and serious visual problems, requiring frequent wiping on the inside of the face piece to clear condensation.
Heat and humidity management is a problem with both level A and B suits.

FIG. 3—*Focus group 1: problems with level A and B suits.*

When subjects were asked "when" cooling was needed, the response was "usually immediately," prompting a discussion about the perception that a level B suit was a tighter fitting garment than level A, and, therefore, heat and moisture build-up were serious problems in level B suits as well. The subjects reported that an ambient temperature above 15.6°C (60°F) can cause heat exhaustion when wearing either suit.

This focus group provided insight into the workings of HAZMAT teams. Respondents reported that there are usually two sets of HAZMAT members at a site. A two-person team goes in to work the incident, while the other team prepares to go in. The team in waiting is in their suits, not completely zipped up, but already getting hot. If they are in level B suits, they can remove the hood and the facemasks. In level A suits, it is not practical to remove the headpiece. The first team comes out before they deplete their air supply and relays relevant information about the situation, and the second team goes in as the first team gets decontaminated. Alternatively, the same team could be sent back in. This is done when their knowledge of the incident can assist them in efficiently and effectively dealing with the situation or if there is an insufficient number of HAZMAT members. In this situation, a quick decontamination would be done, an air bottle changed, possibly water/liquid offered to drink and perhaps to pour on one's head to cool down, before the team would return to work the incident. This procedure takes about 15 min, while the decontamination portion takes only about 2–3 min. This practice saves time in dealing with the incident, but it adds to the workers' heat stress problems.

Decontamination teams typically use soap and water for cleaning, and the process is often repeated using a shower-like device. They have other chemicals at their disposal; however, they do not use them often because they can create an exothermic reaction on the skin and cause thermal burns. Additionally, unanticipated chemical reactions can occur between the chemical contaminants on the suit and the substances used for decontamination. If a substance is difficult to decontaminate, subjects reported that they usually throw the suit away.

The subjects were asked to describe their experience with currently available cooling vests. Several reported that the vests cool initially, but that the cooling diminishes when they really need it. They recommended that the vest be designed with the worst-case scenario in mind. They recommended that the control system be either an on/off system or be adjustable using automatic/electronic core temperature monitoring devices, or voice activated controls. Since heat removal from the cooler system is such a critical issue, subjects were asked to comment on the concept of placing the cooler unit on the outside of the protective garment. They responded that this would cause a variety of problems, including decontamination problems. They indicated a preference for attaching cooler components to the respirator harness or body so that the person changing the bottle could also change a cooler component or adjust cooling if the unit featured an adjustable system.

Overall, the findings indicated that heat stress was a significant problem, and existing cooling garments were not meeting the needs of HAZMAT workers. There is a need for a cooling garment, and it should be washable. The group was helpful in identifying users' concerns regarding the need for simplicity in design of a cooling system and the need to keep the system as lightweight as possible because of the weight of the respiratory system and HAZMAT ensemble. Fogging or condensation that obscures vision is a serious problem.

Focus Group 2

Although the prototype cooling system was not yet developed, the researchers wanted to obtain user reaction to the system concept. Therefore, a mock-up of a 7.5 cm × 15 cm × 20 cm rectangular cooler unit, weighing 3.816 kg, and an existing liquid-cooled vest were used to solicit user feedback from the six male volunteer subjects.. This focus group was pivotal to the project since it was essential to acknowledge that the prototype cooler would likely need to be located outside level A and B suits and to obtain user feedback on this issue. Key findings related to the cooler unit are given in Fig. 4.

Weight of the stimulus unit was acceptable, but lighter would be better.
Point of attachment between PPE and the cooler unit must be secure to minimize movement or weight shift.
Balanced weight distribution was reported as extremely important for safety reasons.
Location of the cooler unit outside the suit was acceptable if concerns can be fully addressed
Most preferred location for the cooler unit was near the air tank (top, sides, or beneath).
Attachment methods between the PPE and the cooler unit using hook and loop tape or a short, partial belt was acceptable.
Use of belts, inside or outside, the HAZMAT suit was not acceptable.
Shape of the cooler should be reconfigured.

FIG. 4—*Focus group 2: responses to cooler issues.*

The second focus group discussion centered primarily on cooler issues, specifically on locating a cooler unit outside level A and B encapsulated suits. Although the participants initially indicated strong concern with this arrangement, they concluded that an exterior location could be acceptable to them if the device provided cooling as promised, and other associated problems with an exterior location could be solved. The associated problems included concerns about survival of the cooler unit if it was used in various hazardous environments and subsequently for decontamination processes, the need for an approved PPE pass-through, and other safety issues. Participants suggested near the air tank as the most workable location for the cooler, and they suggested altering the shape to eliminate sharp corners and to achieve a configuration allowing placement near the air bottle that would not demand greater clearance.

Key findings related to the interface between the garment and the cooler unit included locating the pass-through as near to the cooler unit as possible to eliminate the potential for tubing interfering with work or getting caught on something, and locating the pass-through as far from arm and hand movement as possible. Key findings related to the garment design are given in Fig. 5.

No tubing should be placed in the shoulder area to eliminate discomfort from the harness pressing on the tubing.
Sometimes there is a need to don and doff in public, therefore, some might want to wear the vest over a T-shirt.
Most preferred a snug fitting garment to increase skin contact and minimize shifting.
Most preferred that the vest manifold be located in the upper or lower center back area.

FIG. 5—*Focus group 2: responses to garment issues.*

Focus Groups 3–6

The purpose of the last four focus groups was to obtain user feedback on the alpha prototype garment and cooler system from first responders trained in HAZMAT operations from various cities, and also to determine if the feedback from the first two focus groups was similar to feedback from first responders from a broader geographical representation. Following initial questions taken from the first focus group, the alpha prototype cooler unit and vest were presented for the HAZMAT responders to see and handle before asking questions related to the specifics of the acceptability of the cooling unit and vest. Results were organized into the following themes: 1) general comments related to focus group 1 input; 2) comments directed to the cooling unit, including size, weight, and location; and 3) comments directed to the prototype cooling garment and the interface between the garment and cooling unit. General comments on the need for cooling and experiences with level A and B ensembles and cooling garments are given in Fig. 6.

Subjects reaffirmed the need for a cooling system.
Subjects reaffirmed that level B suits were used more than level A suits.
Three of the four groups reported experience with cooling vests, primarily frozen gel packs, and reported problems with cooling vests.
Subjects expressed interest in a cooler system.

FIG. 6—*Focus groups 3–6: general comments.*

Respondents in the last four focus groups reaffirmed that level B suits were used far more often than level A suits and that there was a need for personal cooling when wearing level A and B suits. Several said cooling was a necessity and that it helped get the job done. Three of the groups, those located in hot climates, had considerable experience wearing various cooling garments over a period of time. The most common garment was a vest with pockets designed for use with frozen inserts (gel or phase change). The vest was worn under the SCBA. The fourth group had experience using cooling garments during training only. Dissatisfaction with existing gel or phase change vests was expressed by those who had experience using these vests. Complaints on logistics (keeping a freezer on the rig and devoting a person to inserting the often misshapen frozen gel packs into the pockets), discomfort with too much cooling initially or in certain body positions, loss of cooling over time, and the extra weight, were expressed. These comments are summarized in Fig. 7.

Packets became irregular in shape if not frozen in a completely flat configuration, thus, they were difficult to insert in the vest pockets.
Frozen packets were uncomfortable under the SCBA equipment; therefore, they were frequently not placed in the pockets where the SCBA frame and harness pressed against the back.
Vests with fewer pockets for inserting the frozen packets provided only cooling in "spots."
Some cooling was considered to be better than no cooling by some responders.
Vests filled with frozen packets were heavy and became dead weight after thawing.
Frozen packets provided maximum cooling at the beginning of the response period rather than later when more body heat needed to be removed.

FIG. 7—*Focus groups 3–6: HAZMAT responders' experiences with cooling vests.*

The second major topic discussed was the alpha prototype cooler and various related issues. A duration of 30 min was reaffirmed, as was a preference for an on/off control system, although one group preferred an automatic sensing system. Considerable debate ensued over the question of placing the cooler unit outside encapsulated suits. The initial reaction of all four focus groups was that the cooler should be contained inside the suit, and that an outside suit location introduced many issues to be resolved, as summarized in Fig. 8.

The cooler must be capable of decontamination, and must withstand acids, corrosives, chemicals, flash fire, soap and water, and dust.
The cooler must be "intrinsically safe" and must be durable, "fire fighter proof."
The cooler must not have sharp edges and must not compromise the suit integrity.
The location of the cooler on the body cannot present a balance problem.
The cooler must not move or shift during work activities and must have an approved PPE pass-through.
The team should consider integrating the cooler with existing PPE.

Respondents indicated that a system weighing 2.270 kg or less would be ideal. They also indicated that volume, shape, and location of the weight were equally important factors. Focus group participants stated that anything that extends into space needed for the responder is a major issue when concerned with negotiating small spaces, walking on irregular surfaces, and trying to maintain balance.

The third major topic discussed in three of the four focus groups was the cooling vest. One group responded to an incident during the focus group and therefore was not available for questions about the vest. Respondents in three of the four focus groups rated the vest positive for the following features: lightweight, washable, easy to don and doff, fit adjustment mechanism and closure method, quick connect and disconnect with the cooler unit, and appears to fit the body closely. Improvement in manifold design and placement were suggested. Several said the vest looked more workable than what they were using, and most indicated a willingness to test the next generation system. They reaffirmed that those working decontamination could benefit from this system, and they suggested considering the system for rehabilitation uses as well.

Conclusion

A series of six focus groups was held with first responders in five cities in the northeast, midwest, south, and southwest areas of the U.S., as part of the process to develop a personal, portable cooling system for first responders wearing level A and B ensembles. All focus groups stated that the need for reducing heat stress was not limited to level A encapsulated suits. All stated that the level A HAZMAT suits were worn from zero to less than five times a year, typically. However, level B suits were used hundreds of times a year in responding to a variety of incidents. All confirmed the problem with heat and humidity buildup in the encapsulated suits and the associated problem of diminished visibility. All groups stated a resounding "yes" when asked if there was a need and willingness to use a cooling system if problems associated with currently available vests could be solved. Major concerns related to having a system that would be quick and easy to don and doff, that would not interfere with work to be accomplished, and that did not compromise the safety of the HAZMAT responder. Respondents' information on the kinds of physical and cognitive activities typically done in level A and B suits was used to design a work protocol for a subsequent physiological study, as indicated in Fig. 1.

In summary, the focus group methodology generated valuable information for the research team to improve the alpha prototype cooling system, to target the system to appropriate user groups, and to plan subsequent components of the 3-year study.

Acknowledgments

This research is supported under Award MIPT-106-113-2000-021 from the National Memorial Institute for the Prevention of Terrorism (MIPT) and the Office for Domestic Preparedness, U.S. Department of Homeland Security. Points of view in this document are those of the authors and do not necessarily represent the official position of the U.S. Department of Homeland Security or of MIPT. The authors thank William Haskell of NIOSH-National Institute of Occupational Safety and Health, National Personal Protective Technology Laboratory; Chief Bryan Hierston of the Oklahoma City Fire Service; and Professor James Hanson of Oklahoma State University, for their assistance in obtaining cooperation from the HAZMAT responder community to participate in this study.

References

[1] Press Release (retrieved November 2003). Supporting First Responders Strengthening Homeland Security, http://www.dhs.gov/dhspublic/display?content=119, para. 1 & 2.

[2] Hohl, B. "Selecting the Right Protective Clothing," *Plant Engineering*, July 2001, pp. 64–68.

[3] Ruhl, C. "EPA Region 6: Risk Management Programs," *Seventh Annual Fire Protection and Safety Symposium*, pp. 1–15.

[4] EPA Levels of Protection (retrieved November, 2003). Occupational Safety and Health Administration, http://www.osha.gov/dts/osta/otm/otm_viii/otm_viii_1.html.

[5] Touliatos, J., and Compton, N. H., *Research Methods*, Ames, IA: Iowa State University Press, 1988.

Journal of ASTM International, February 2005, Vol. 2, No. 2
Paper ID JAI12103
Available online at www.astm.org

Kevin M Semeniuk,[1] *Jean-Philippe Dionne,*[1] *Aris Makris,*[1] *Thomas E Bernard,*[2] *Candi D Ashley,*[2] *and Theresa Medina*[2]

Evaluating the Physiological Performance of a Liquid Cooling Garment Used to Control Heat Stress in Hazmat Protective Ensembles

ABSTRACT: The purpose of this study was to evaluate the physiological performance of a liquid cooling garment in controlling heat stress in individuals wearing Hazmat protective ensembles. Five heat-acclimated male subjects completed three trials each: standard work clothing (WC), Hazmat suit (H), and a Hazmat suit with torso cooling (HC). All testing took place in an environmental chamber (35°C and 50 % relative humidity) with subjects walking on a treadmill at a metabolic rate of 300 W. Trials were performed until subjects completed the maximum 90 min, achieved the predetermined endpoints in core temperature (38.5°C) or heart rate (95 % of age predicted maximum), or symptoms of fatigue, disorientation, and discomfort forced the participant to discontinue or the researchers to end the test. A significant difference ($P<0.05$) between HC and H was found in final heart rate and core temperature (T_{re}), rate of change in heart rate and T_{re}, and exposure time. As well, HC was similar to WC in all measured physiological variables. It was concluded that a liquid cooling vest was beneficial in alleviating the heat stress caused by Hazmat protective ensembles.

KEYWORDS: liquid cooling garment, vapor permeability, Hazmat, heat stress, thermoregulatory response, core temperature

Introduction

Vapor impermeable ensembles, such as Hazmat or Nuclear, Biological, and Chemical (NBC) suits, are an integral component for providing protection against gas and liquid contaminants. Unfortunately, with the decreased diffusion of water vapor, an extremely humid microclimate is created that prevents sweat evaporation and thus, the body's only active means to transfer heat to the surrounding environment. In examining the thermoregulatory response of individuals wearing various chemical and biological protective ensembles, researchers found that all suits imposed significant heat strain during 100 min of exercise, at approximately 400 W, in an environmental chamber at 35°C and 50 % relative humidity [1]. Compounding the heat stress is necessary because individuals requiring vapor protection routinely work in environments with hot temperatures and high humidity. For instance, workers involved in decommissioning and decontaminating nuclear facilities with the Department of Energy, are exposed to ambient

Manuscript received 1 December 2003; accepted for publication 15 July 2004; published February 2005. Presented at ASTM Symposium on Performance of Protective Clothing: Global Needs and Emerging Markets: 8th Symposium on 13 January 2004 in Tampa, FL.

[1] Research and Development Department, Med-Eng Systems Inc., 2400 St. Laurent Blvd., Ottawa, ON, Canada, K1G 6C4.
[2] College of Public Health, University of South Florida, 13201 Bruce B. Downs Blvd., Tampa FL 33612-3805.

temperatures as high as 40.5°C [2]. As a consequence to the occupational requirements, the likelihood of metabolically and environmentally induced heat stress is greatly increased.

Due to the reduced vapor permeability of totally encapsulating clothing ensembles (e.g., Hazmat ensembles), the body's thermoregulatory mechanisms become impaired and unable to maintain an acceptable body temperature, potentially leading to the occurrence of heat-induced disorders such as fatigue, cramps, dehydration, and exhaustion. These symptoms would inevitably affect job performance and individual health and safety. As a result, a number of institutions have produced material and outlined specific administrative controls regarding work in hot environments, with the goal of improving job production and worker safety. The Occupational Safety and Health Administration (OSHA) provides information concerning heat disorders and health effects, while discussing sampling methods and control procedures [3]. The World Health Organization recognizes a maximum core temperature of 39°C for intermittent exposures, but recommends 38°C for continuous exposures to heat stress. The American Conference of Governmental Industrial Hygienists (ACGIH) proposed threshold limit values based on metabolic demands and the wet bulb globe temperature (WBGT) index to evaluate heat stress with various protective ensembles, in order to specify safe working environments that could be tolerated for about eight hours [4,5].

In an industry utilizing protective clothing, heat stress control should be of paramount importance from a health and safety perspective. However, work requiring vapor impermeable clothing is necessary and will always be undertaken regardless of the environmental conditions and the various controls taken to minimize the risk. Therefore, technology must be developed to protect the individual workers from the risk of heat stress, while maintaining production and cost effectiveness. The development and use of a Personal Cooling System (PCS) provides a solution to heat stress management. A PCS works to primarily remove internally generated metabolic heat and secondarily block external environmental heat.

Numerous studies have examined the use of various PCSs, and in particular Liquid Cooling Garments (LCGs), in alleviating heat stress when wearing protective clothing in hot environments [2,6,7,8]. An LCG, by conductive heat transfer, cools the body with circulating water usually traveling through a network of tubing within the garment. One such study found that the LCG was a viable means of controlling thermal strain, by decreasing core temperature, in engineering personnel wearing chemical defense clothing aboard Navy ships [9]. As well, the U.S. Navy Clothing and Textile Research Facility (NCTRF) conducted laboratory evaluations of a prototype microclimate cooling system designed for use with chemical protective clothing [10]. Use of the system reduced rectal temperature, and hence the physiological heat strain, with the impermeable ensemble. Other studies have compared the performance of various PCSs in reducing the level of heat stress in protective ensembles. One such study demonstrated that an LCG that circulates chilled water through a tubing network successfully reduced heat stress, as compared to other cooling systems in subjects exercising in Hazmat suits in an environmental chamber (35°C and 50 % relative humidity) [11].

Previous research, however, has largely focused on PCS projected for military purposes. As well, the LCG investigated tended to be full body suits, where water temperature and flow were controlled and supplied by large chillers. Instead, research should also focus on PCS systems that are appropriate for industrial applications where the majority of individuals interested in personal cooling require a system that will maintain an individual's thermal steady state, be easily donned and doffed, be capable of being worn underneath protective clothing, and is controlled by a lightweight, portable unit.

The purpose of this study was to examine the performance of an industrial applicable LCG in controlling heat stress with Hazmat protective clothing ensembles. Physiological measurements were derived during three conditions: standard work clothing (WC), Hazmat suit (H), and Hazmat suit with cooling (HC). Heat-acclimated subjects were used to avoid partial heat acclimation and simulate acclimatized industrial workers. Subjects exercised for a maximum period of 90 min in an environmental chamber. It was hypothesized that the use of personal cooling worn underneath a Hazmat ensemble would reduce the effects of heat stress to levels similar with standard work clothing.

Methods

Subjects

Five healthy males participated in this study. The average (±SD) age, height, and weight was 32.6 ± 10.8 years, 179.6 ± 4.0 cm, and 92.3 ± 9.4 kg, respectively. Prior to testing, all subjects had to undergo a medical examination by a qualified physician and give their voluntary and informed consent to participate in this experiment. None of the subjects had ever experienced a previous occurrence of heat stroke. This research was approved by the Institutional Review Board of the University of South Florida according to the guidelines of the National Institutes of Health to ensure subject safety.

Experimental Protocol

Prior to experimental testing, participants completed a heat acclimation protocol. Subjects were required to walk on a treadmill on five consecutive days in an environmental chamber at 50°C and 20 % relative humidity while wearing athletic shorts, shoes, socks, and a T-shirt. The maximum time allotted for each heat acclimation session was 2 h. The sessions were conducted at approximately the same time each day, and subjects were allowed to drink ad libitum. The protocol also allowed researchers to determine an initial treadmill velocity that corresponded to the required metabolic work rate.

The experimental protocol consisted of 3 trials: standard work clothing (8 oz cotton pants and 6 oz cotton work shirt with a thermal insulation of approximately 0.6 clo), Hazmat suit (a Level B polyethylene-coated DuPont Tychem® QC coverall with hood), and Hazmat suit with a liquid cooling garment (Med-Eng Systems CardioCOOL™ and PortaCOOL™). All garments were sized to the subjects based on the manufacturer's instructions. For example, an LCG was chosen to be tight-fitting yet not restrictive. Participants wore athletic shorts, socks, shoes, and a T-shirt underneath all three ensembles. A respirator was not used with the Hazmat ensemble due to the interference with the equipment used to measure metabolic cost. Chilled water was supplied to the LCG through closed circuits of small plastic tubing sewn into the garment. A vest design was chosen as the LCG (CardioCOOL™) because of its versatility and relevance to industrial applications. As mentioned, the vest was worn tight to the skin to maximize the conductive heat transfer. The chilled water was generated by the portable cooling unit (PortaCOOL™), which used a pre-frozen 2 L ice bottle as the heat sink. A battery powered pump circulated the chilled water through the tubing at a flow rate of approximately 350 ml/min, where heat was extracted from the body and then exchanged directly with the ice bottle upon return to the cooling unit. Batteries were replaced after 4.5 h of use to ensure that flow rate was maintained. For the purpose of these tests, the cooling unit was worn outside the Hazmat suit around the waist to facilitate easy replacement of the ice bottles, which were exchanged before the ice was fully exhausted. A maximum of three bottles was required for any one subject. The total weight of the cooling system charged with water and including the LCG was approximately 7 kg.

The environmental chamber was maintained at 35°C and 50 % relative humidity. Air motion was kept at 0.5 m/s, and the dry bulb and wet bulb temperatures were held constant within 0.5°C. In this environment, participants attempted a 90-min treadmill walk at a metabolic rate of approximately 300 W (Fig. 1). The trial order was counterbalanced across subjects. A trial was terminated when a subject successfully completed the entire 90 minutes. If a subject achieved the predetermined endpoint criteria for core temperature (38.5°C) or heart rate (95 % of age predicted maximum) or physiological symptoms of fatigue, impending syncope, headache, and/or disorientation, the subject was forced to discontinue, or the researchers ended the test.

Physiological measurements were made throughout the entire experimental protocol. Initial recordings began when the participant was in the chamber and started to exercise on the treadmill. A heart rate monitor (Polar) was used to measure heart rate. Core temperature (T_{re}) was measured in the rectum with a YSI model 401 thermistor, which was inserted 10 cm beyond the anal sphincter. Heart rate and T_{re} were recorded every 5 min. The overall rate of rise in both heart rate and T_{re} were determined using the initial and final values and the total exposure time. Metabolism during exercise was measured at every 15-min interval. Metabolic rate was determined using an open circuit indirect method with expired gases being collected with a Douglas air bag (Fig. 1).

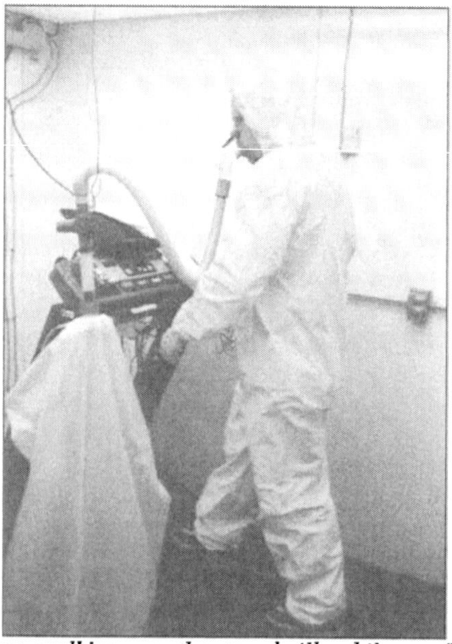

FIG. 1—*Subject shown walking on the treadmill while performing Hazmat condition protocol. A Douglas air bag is shown to collect expired gases for the determination of metabolic rate.*

Statistical Analysis

Physiological responses due to heat stress were analyzed using a simple repeated measure analysis of variance to determine if any significant differences emerged as a result of the various clothing conditions. Tukey's HSD post hoc test was performed when statistical differences between means were found ($P<0.05$).

Results

Descriptive statistics, represented by averages and standard deviations, were calculated for all measured physiological variables (Table 1). As expected, no significant differences were found in average metabolic rate between clothing conditions, which indicates an equivalent work input in all conditions. In fact, average metabolic rate differed by less than 5 % between any of the conditions. In the WC and HC conditions, each subject completed the entire 90 min of the experimental protocol, whereas in condition H, only two subjects were able to exercise for the total 90 min. Of those three subjects who did not complete the protocol, one was terminated due to maximum T_{re} being attained (50 min) and two because of exhaustion from heat strain (50 and 55 min). Average exposure time for WC, HC, and H conditions 90, 90, and 67 min, respectively (Fig. 2). A significant difference was found between H and HC, and H and WC for exposure time ($P<0.05$).

The average final heart rate was found to be significantly higher ($P<0.05$) for the H condition, 145 ± 10.4 bpm, as compared to both the WC and HC conditions, 94.6 ± 2.7 and 103 ± 13.8 bpm, respectively (Fig. 3). No significant difference was found between the WC and HC conditions. Similar results were also determined in the difference between initial and final heart rate and rate of change of heart rate, with no significant differences derived between WC and HC, but where both conditions differed significantly from H ($P<0.05$). No significant differences were found in initial heart rate between the three conditions.

TABLE 1—*Descriptive statistics (averages ± standard deviations) for measured parameters for the three clothing conditions: standard work clothing (WC), Hazmat suit (H), and Hazmat suit with cooling (HC).*

Variable	Standard Work Clothing	Hazmat	Hazmat and Cooling
Metabolic Rate, W	317 ± 38.3	323 ± 28.5	332 ± 48.5
Walking Velocity, m/s	0.92 ± 0.19	0.85 ± 0.19	0.85 ± 0.19
Exposure Time, min	90.0 ± 0.0	67.0 ± 21.1	90.0 ± 0.0
Initial Heart Rate, bpm	85.6 ± 3.5	91.0 ± 12.4	86.6 ± 6.0
Final Heart Rate, bpm	94.6 ± 2.7	144.6 ± 10.4	102.6 ± 13.8
Final – Initial Heart Rate, bpm	9.0 ± 4.1	53.6 ± 14.5	16.0 ± 11.0
Rate of Change in Heart Rate, bpm/min	0.100 ± 0.046	0.839 ± 0.214	0.178 ± 0.122
Initial T_{re}, °C	37.26 ± 0.363	37.31 ± 0.356	37.14 ± 0.154
Final T_{re}, °C	37.68 ± 0.159	38.43 ± 0.211	37.95 ± 0.296
Final – Initial T_{re}, °C	0.42 ± 0.215	1.12 ± 0.318	0.81 ± 0.306
Rate of Change in T_{re}, °C/min	0.005 ± 0.002	0.017 ± 0.004	0.009 ± 0.003

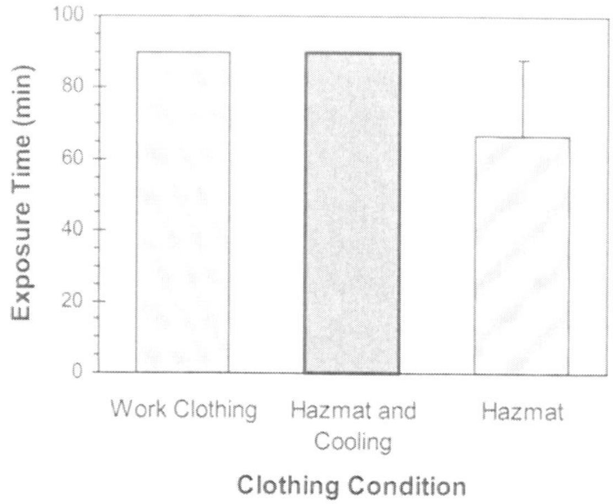

FIG. 2—*Average exposure times for each of the three clothing conditions; standard work clothing (WC), Hazmat suit with cooling (HC), and Hazmat suit (H).*

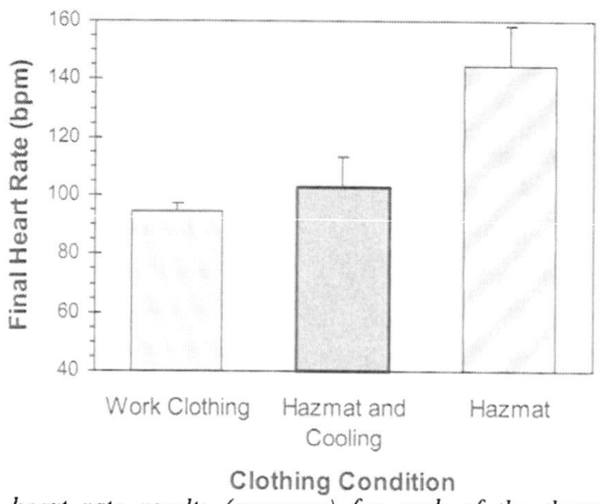

FIG. 3—*Final heart rate results (averages) for each of the three clothing conditions; standard work clothing (WC), Hazmat suit with cooling (HC), and Hazmat suit (H).*

An overall significant difference was also found among all three conditions when assessing final core temperature, whereas no significant difference was found between conditions when examining initial core temperature ($P<0.05$). Final T_{re} measured in HC (37.95 ± 0.30°C) and WC (37.68 ± 0.16°C) were significantly different from condition H (38.43 ± 0.21°C) (Fig. 4). No difference in final T_{re} was found again between HC and WC. The rate of change in T_{re} also produced similar trends, in that a significant difference ($P<0.05$) was found between HC and H, and WC and H, but not between HC and WC.

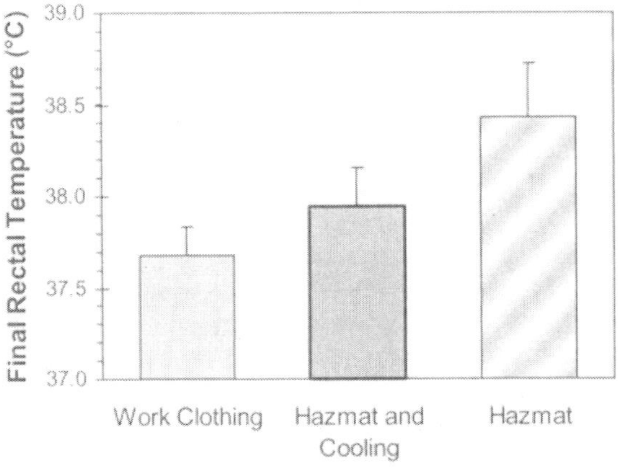

FIG. 4—*Final core temperature results, as determined by rectal temperature (T_{re}), for each of the three clothing conditions: standard work clothing (WC), Hazmat suit with cooling (HC), and Hazmat suit (H).*

Discussion

This study described the physiological responses to heat stress of a vapor impermeable protective ensemble worn with and without personal cooling as compared to standard work clothing. A portable liquid cooling system was examined as a result of previous research assessing performance [2,6,7,9,12,13], the thermal properties of a water and ice based system, its ease of use, and its applications to industry. The exercise rate was chosen to represent the metabolic cost of typical occupational tasks. Although more difficult to control, work intensity was determined using the metabolic rate measured during testing as compared to indicating an external work rate. Therefore, non-significant variations were evident between averages with the metabolic rate being higher in HC (332 W) than in WC (317 W) and H (323 W). The increased metabolic requirement in HC was due to the additional mass of the LCG control unit (7 kg) the participant was required to wear during the cooling condition. As a result, average walking velocity was also similar between HC and H, and as expected, slightly lower than WC.

Overall, this study found that the LCG, when worn underneath a Hazmat ensemble, positively affected exposure time, exercise heart rate, and core temperature response. Comparable results were found in a study at the U.S. Navy Clothing and Textile Research Facility (NCTRF) assessing a prototype microclimate cooling system (MCS) with two different chemical protective ensembles [10]. In this study, seven male subjects attempted 120 min of heat exposure in an environment of 35°C with 60 % relative humidity while exercising at approximately 300 W. As in the present case, a maximum exposure time occurred when cooling was provided. Use of the MCS reduced rectal temperature by an average of 0.5°C after 120 min with the semi-permeable ensemble and by 1.3°C after 100 min with the impermeable ensemble. The MCS also reduced heart rate with the semi-impermeable and impermeable ensembles, 30 and 42 bpm, respectively. In the present study, the LCG equally reduced rectal temperature by 0.5°C and heart rate by 40 bpm after 90 min of exercise. It is evident that if not controlled, vapor impermeable protective ensembles can create a hot and humid microclimate within the suit by affecting evaporative cooling used to transfer heat to the environment. Incorporating the LCG altered the microclimate within the Hazmat ensemble so that clothing characteristics were similar, physiologically, to standard work clothing in an identical environment, by reducing the strain on the body's thermoregulatory mechanisms.

Particularly interesting was the performance of the vest, which was incorporated into the protocol due to the increased relevance to industrial applications. Significant differences between Hazmat conditions were still obtained by providing cooling to only the torso instead of the whole body. Workers require a personal cooling system that is lightweight, easy to use, and portable; requirements attained by the chosen LCG. Furthermore, acclimating subjects in this study more closely represented actual Hazmat users who are constantly exposed to various levels of heat stress, as when wearing vapor impermeable ensembles. Adaptations following heat acclimation in typical work clothing could include favorable decreases in heart rate, core temperature, skin temperature, and perceived rate of exertion when exposed to heat stress. However, conflicting results arise when examining the benefits of acclimation and the use of protective clothing. It has been determined that after eight weeks of aerobic training or six days of heat acclimation, little or no improvements in physiological responses or tolerance time were observed in subjects wearing NBC protective ensembles [6]. It was thought that these results might be due to the limited water vapor permeability through the clothing, where it was possible that the increased sweat production in trained or heat-acclimated individuals may have increased physiological strain by promoting a faster rate of dehydration rather than increased evaporative heat loss. Therefore, providing an external means of cooling can be incorporated to improve health and safety since acclimatization potentially does not alleviate the effects of heat stress for individuals wearing relatively impermeable protective clothing.

In summary, a liquid cooling garment proved beneficial in controlling heat stress in Hazmat protective ensembles. Most of the variables measured – exposure time, heart rate, and core temperature – indicated that the LCG, when worn underneath the Hazmat suit, reduced physiological strain to levels comparable to standard work clothing. In fact, no significant differences were found between any of the WC and HC physiological variables, which indicates a major advantage in wearing a liquid cooling system than the Hazmat suit alone. Given that heat stress inevitably decreases performance and leads to fatigue, maintaining a proper thermal balance with a personal cooling system may increase individual health and safety.

References

[1] Levine, L., Johnson, R. F., Teal, W. B., Merullo, D. J., Cadarette, B. S., Staab, J. E., et al., "Heat Strain Evaluation of Chemical Protective Garments," *Aviation Space and Environmental Medicine*, Vol. 72, No. 4, 2001, pp. 329–335.

[2] Dobbs, K. and Prochaska, M., "Successful Introduction of the PICS Technology to the DOE Weapons Complex," *US Department of Energy*, DOE/EM-0393, 1998.

[3] Occupational Safety and Health Administration, "OSHA Technical Manual: Heat Stress," U.S. Department of Labor, 2001.

[4] American Conference of Governmental Industrial Hygienists, "Threshold Limit Values for Chemical Substances and Physical Agents," 2003.

[5] Kenney, W. L., "WBGT Adjustments for Protective Clothing," *American Industrial Hygiene Association Journal*, Vol. 48, 1987, pp. A576–A577.

[6] Aoyagi, Y., McLellan, T. M., and Shepard, R. J., "Effects of Training and Acclimation on Heat Tolerance in Exercising Men Wearing Protective Clothing," *European Journal of Applied Physiology*, Vol. 68, 1994, pp. 234–245.

[7] Frim, J. and Morris, A., "Evaluation of Personal Cooling Systems in Conjunction with Explosive Ordnance Disposal Suits," *Defense and Civil Institute of Environmental Medicine*, DCIEM No. 92-31, 1992.

[8] Moutain, S. J., Sawka, M. N., Cadarette, B. S., Quigley, M. D., and McKay, J. M., "Physiological Tolerance to Uncompensable Heat Stress: Effects of Exercise Intensity, Protective Clothing, and Climate," *Journal of Applied Physiology*, Vol. 77, No. 1, 1994, pp. 216–222.

[9] Frim, J. and Glass, K., "Alleviation of Thermal Strain in Engineering Space Personnel Aboard Ships with the Exotemp Personal Cooling System," Defense and Civil Institute of Environmental Medicine, DCIEM No. 91-62, 1991.

[10] Pimental, N., Teal, W., and Avellini, B., "Effectiveness of a Prototype Microclimate Cooling System for Use with Chemical Protective Clothing," US Navy (NCTRF), Technical Report No. 180, 1990.

[11] Bernard, T. E., Hart, S. D., and Richards, I. S., "Physiological Evaluation of Core-Control Systems: A Comparison of a New High-Performance Whole-Body Cooling System to Other Personal Cooling Systems and a Control," College of Public Health, University of South Florida, 1992.

[12] Jetté, F. X., Dionne, J. P., Semeniuk, K .M., and Makris, A., "Thermal Manikin Investigation of Personal Cooling Systems Worn under Protective Clothing with Various Permeability and Insulation Properties," *2003 SAFE Symposium*, SAFE Association, Creswell, OR.

[13] Dionne, J. P., Semeniuk, K. M., Makris, A., Teal, W., and Laprise, B., "Thermal Manikin Evaluation of Liquid Cooling Garments Intended for Use in Hazardous Waste Management," *2003 Waste Management Symposium*, WM Symposium Inc., Tucson, AZ.

SECTION 3: PREDICTING PRODUCT PERFORMANCE

Journal of ASTM International, May 2004, Vol. 1, No. 5
Paper ID JAI12121
Available online at www.astm.org

Helena Mäkinen,[1] Sanna Karkkula,[1] Kirsi Jaakkola,[1] and Patrick von Nandelstadh[1]

Can Laboratory Pre-Treatment Of Welders' Protective Clothing Simulate Their Aging Process In Normal Use?

ABSTRACT: The fabrics of welders' protective garments can be worn out after only one or two use-wash cycles. It is probable that the by-products in the welding process, i.e., ultraviolet C-type (UVC) radiation, ozone, metal fumes, can contribute to the molecular changes responsible for the degradation. In this study we simulated the normal use of garments by cycling UVC/ozone and washing pre-treatments in the laboratory. In parallel, garments of the same five study fabrics were used by welders in a shipyard until the end of their service life. Properties defined in prEN ISO 11611were measured from the laboratory-treated and the normally used fabrics. The laboratory cycling treatment affected tear resistance more than the tensile strength of the study fabrics. Both pre-treatments seemed to have affected the flammability of the fabrics similarly. The results of this study support the conception that it is possible to simulate the aging process of normally used garments by laboratory pre-treatments.

KEYWORDS: fabric, UVC, ozone, welding, pre-treatment, degradation, tensile strength, tear strength, flame spread

Introduction

In heavy welding processes, the fabrics of welders' protective clothing have a tendency to become brittle in a short period of wearing time. The garments may be worn out after only one or two use-wash cycles. It is probable that the by-products in the welding process, i.e., UVC radiation, ozone, metal fumes, can contribute to the molecular changes responsible for the degradation. If this degradation proceeds without visible signs in the clothing, the user is at risk, e.g., if the flammability properties have fallen to a dangerous level.

The arcs associated with arc welding emit high levels of ultraviolet radiation (UVR). UVR is divided into three bands, the wavelength ranges and common names of which are [1]:

- UV-A: 315-400 nm
- UV-B: 280-315 nm
- UV-C: 100-280 nm

High irradiance levels take place in the UV-C ranges during welding. Figure 1 shows a luminous spectrum of MIG welding on aluminum [2]. This short-wave UV radiation produces

[1]Senior Researcher, Research Engineer, Assistant Researcher, Laboratory Engineer, respectively, Department of Physics, Finnish Institute of Occupational Health, Topeliuksenkatu 41 aA, FIN-00250 Helsinki.

Presented at: Performance of Protective Clothing: Global Needs and Emerging Markets: Eighth Symposium, January 13, 2004, Tampa, FL.

ozone (O_3) gas, which can also cause health hazards. Ozone is produced in arc welding by a photochemical reaction when short-wave 130–200 nm UV radiation affects the oxygen of the air. It is an unstable gas with a strong and irritating odor, and is also highly reactive. MIG and TIG welding result in the highest ozone levels. The ozone concentration increases depending on the type of electrodes used, amperage, extension of arc time, and increased argon flow. If welding is carried out in confined and poorly ventilated spaces, the ozone concentration may rise to harmful levels [2,3].

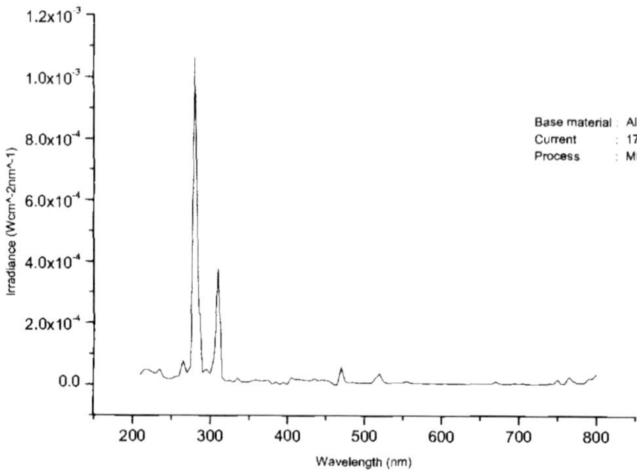

FIG 1—*Luminous spectrum of MIG welding on aluminum* [2].

UVR damages not only the eyes, but also the skin. Therefore, skin protection of welders is very important. Protective clothing protects welders against UV radiation, welding sparks and burns. It should not transmit UVR, and should not be degraded by the effects of the welding environment. Sliney and co-workers [4] found that most industrial clothing fabrics provide adequate protection against UV radiation. This is true if the clothing has not been damaged by sparks and droplets or by other environmental factors.

In addition to the UVR and ozone, the welding environment contains metal fumes, fluorides, carbon monoxide, nitrogen dioxide and nitric oxide. The metal fumes consist of minute particles that are formed when the vaporized metal rapidly condenses in air. The fumes can contain aluminum, cadmium, chromium, copper, iron oxide fume, lead, magnesium oxide fume, nickel, silicon tin, zinc oxide fume, etc. [3,5].

It is expected that the UVR levels during welding work will become higher in the future, as there is a tendency to use high welding currents to increase efficiency. UVR might be hazardous even at a distance of 10 m from the arc. [6]. Tencate and Collins estimated the exposures at clothing surface for welders around 3000 times and for non-welders around 13 times of the maximum permissible exposure limit [7].

According to Hietanen and co-workers [2] the factor in MIG welding which has the greatest influence on the ozone level was the composition of the filler wire. A filler wire containing magnesium emits significantly lower ozone levels than a filler wire containing silicon. The ozone levels in the welder's breathing zone in actual working environments occasionally exceeded the Finnish occupational exposure levels (0.2 ppm) weighted for a fifteen-minute reference period.

Ozone represents a hidden danger to many materials and components and can prematurely age them. A standard test method has been developed to test the resistance of rubber to ozone cracking [8]. The effect of ozone on textile materials has not been studied.

The problem of aging has been taken into account in the European Directive on personal protective equipment (PPE) [9]. If it is known that the design performance of new PPE may significantly be affected by aging, the Directive obligates the manufacturer to inform users about it. Manufacturers of welders' protective clothing have difficulties in informing users about the aging because of the lack of knowledge about the aging of fabrics caused by UVC radiation and ozone. The current European standard for welders' protective clothing requires only 5 washings before testing and evaluation, if the intended fabric meets the requirement [10]. This is obviously not enough to simulate the aging process in normal use.

There are some earlier studies on the effect of UVC radiation. Schlüter and co-workers [11] in Germany found that the fabrics of the garments normally used by welders did not transmit UV radiation, and they found no differences in strength values after UVC pre-treatment, compared to new fabrics. According to Mäkinen and co-workers [12], UVC pre-treatment was more effective in decreasing tear strength than tensile strength, but the aging effect of laboratory pre-treatment was much lower than in normal use.

According to a study of Toyo [13], aramid fiber had the best properties; even after 200 hours of UV irradiation it retained 70% of its original strength, whereas cotton and the tetrose fiber cotton blend retained less than 50%. The author explains these finding to be due to the molecular coupling of the aramid fiber's benzene ring, which creates a strong structure for the material.

If the aging process could be simulated in the laboratory, manufacturers could select materials that best resist the welding environment; they could also develop new durable materials faster. This study compares the aging properties of different types of fabrics used for welders' protective clothing after controlled laboratory pre-treatment by cycling UVC/ozone exposure and washing and normal use in a shipyard.

Materials and Experimental Procedures

The fabrics for the study were chosen to represent traditional and newly developed fabrics, which had been tested and approved in accordance with EN 470-1:1995, Protective Clothing for Use in Welding and Allied Processes [10]. A list of the tested fabrics is shown in Table 1.

TABLE 1—*Description of the materials.*

Fabric No.	Composition and construction	Weight (g/m^2)
1	FR 100% CO, satin	320
2	50/50% CO/PES, satin	340
3	75/25% CO/Kevlar, twill	310
4	55/45% PPAN/CO, twill	325
5	35/65% Conex PBI/ FR CV, satin	310

Pre-treatment

Aging in Normal Use in Welding Work

The garments of shipyard welders who worked in hull production were used to study the normal aging of fabrics in an actual working environment. The working methods and conditions in the shipyard environment were inquired by a questionnaire. The welders used mainly MIG/MAG welding methods.

Six welders wore protective overalls made from the study fabrics. In total, 30 overalls were manufactured. The overalls were cleaned together with the normal overalls of the shipyard in an industrial laundry. The researcher and the stock keeper of the shipyard, who normally evaluated the garments, examined each overall separately after each washing and decided whether the overall was at the end of its service life. At the end of service life, the overalls were taken to the laboratory for testing. The average length of service life of the overalls (working hours/overall) is shown in Table 2. Some overalls were worn out after only one use-wash cycle, while some lasted a maximum of three use-wash cycles. This pre-treatment is later called USED.

TABLE 2—*Service life of overalls.*

Fabric no.	Service life/overall, h	No. of use-wash cycles
1	105	max. 2
2	86	max. 2
3	104	max. 3
4	110	max. 2
5	78	max. 1

Simulated Aging in the Laboratory, Cycling of UVC/ozone and Washing

In the laboratory, the study fabrics were exposed to cycling treatment consisting of UVC/ozone chamber treatment and washing in water 10 times. The fabric samples were exposed to UV radiation emitted from a fused silica envelope high-pressure Xenon lamp (450 W). The lamp emitted UV radiation with an appreciable amount of UVC radiation, which simulates the UVC radiation in normal welding processes. The distance from the lamp to the sample was 300 mm. The samples rotated around the lamp at a speed of 10 cycles/min. When lighted, the Xenon lamp also produced ozone gas. The measured average level of ozone during the exposure time was 4 ± 2 ppm. The exposure time for one cycle was eight hours. To make sure that the exposure was even, the samples were turned regularly after every cycle. After eight hours of UVC/ozone exposure, the fabrics were washed, dried, and hot-pressed according to the manufacturers' instructions and ISO 6330:2000 [14]. This pre-treatment is later called LABCYCLE.

Test and Evaluation Methods

The pre-treated fabrics, USED and LABCYCLE, were measured for the properties as defined in prEN ISO 11611:2002, August, Protective clothing for use in welding and allied processes [15] (See Table 3). The test methods were in accordance to the testing standards defined in prEN ISO 11611 (See Table 3). The test specimens from the used overalls were cut from the front side of the overall, avoiding the most damaged areas. The LABCYCLE specimens were from the treated fabric samples.

All specimens were conditioned for at least 24 h in an atmosphere with a temperature of 20 ± 2°C and a relative humidity of 65 ± 5% before testing. For the electrical resistance test, the specimens were conditioned for at least 24 h in an atmosphere with a temperature of 23 ± 1°C and a relative humidity of 25 ± 3%.

The effect of the pre-treatments on the fabric fibers was studied microscopically (magnification of 1 000 x).

TABLE 3—*Properties measured and test methods.*

Property	Test method	Test parameters
Tensile strength	ISO 13934-1:1999	Jog speed 100 mm/min Gauge length 200 mm Pretension 5 N
Tear strength	ISO 4674:1977, method A1	Jog speed 100 mm/min Gauge length 100 mm
Limited flame spread	ISO 15025:1994, method A	A flame of 15 mm produced by a propane gas burner Ignition time 10 s
Heat transfer (radiation)	ISO 6942:2002	Source of radiation: six silicon carbide heating rods Heat flux density 20 kW/m^2
Electrical resistance	EN 1149-2:1995	Test voltage 100 W Determination after 15 s

Results

Tensile Strength

The mean value of the maximum force of five specimens, both weft and warp directions, is shown in Fig. 2. Maximum force is the force needed the break the test specimen. The minimum value for tensile strength according to prEN ISO 11611 is 400 N. The measured values are over this limit value.

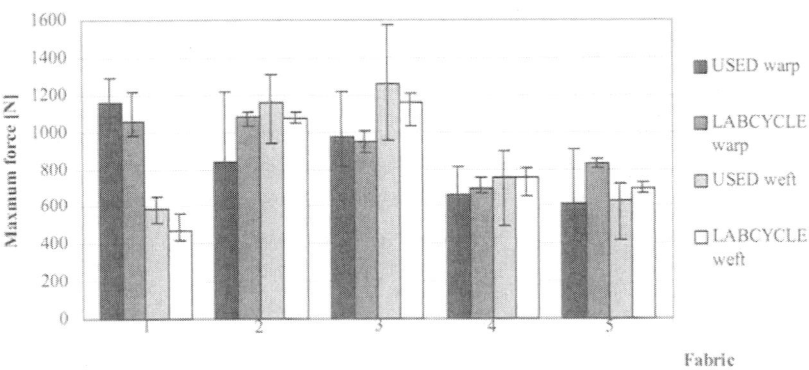

FIG. 2—*Tensile strength of fabrics after USED and LABCYCLE pre-treatment.*

Tear Strength

The mean value of the median force of five specimens, both weft and warp directions, is shown in Fig. 3. Median force is the median of five highest forces on the central 50% of the test specimen. The minimum value for tear strength according to prEN ISO 11611 is 20 N. The measured values are over this limit value.

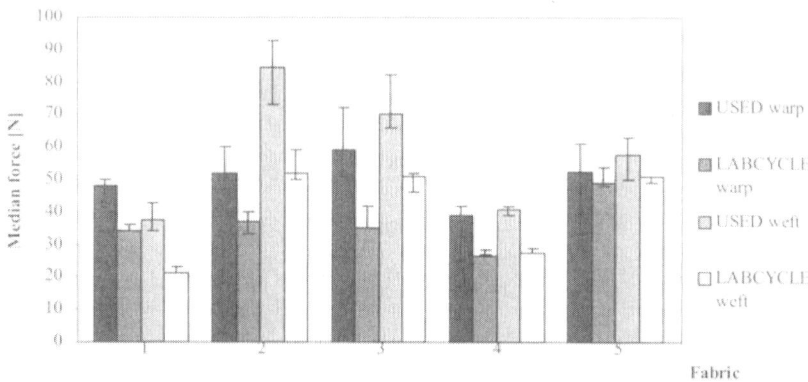

FIG. 3—*Tear strength of fabrics after USED and LABCYCLE pre-treatments.*

Limited flame spread, surface ignition

The results of the flame spread test are shown in Tables 4 and 5. Afterflame and afterglow times are mean values. Afterflame time and afterglow time are the times during which a specimen continues to flame or glow, after the ignition source has been removed.

prEN ISO 11611 states the requirements for flame spread as follows: "afterflame time ≤ 2 s, afterglow time ≤ 2 s, no specimen shall give flaming to the top or either side edge, or shall have hole formation or shall give flaming or molten debris." The recorded observation of fabrics 3, 4, and 5 as USED and LABCYCLE fulfilled the requirement in prEN ISO 11611. For fabric 1 an afterflame time of over two seconds was measured after both pre-treatments, and three LABCYCLE specimens of this fabric flamed to the top edge of the specimen. Also, two specimens of fabric 2, as USED, flamed to the top edge of the specimen, and an afterflame time of over two seconds was recorded.

TABLE 4—*Results of flame spread test for USED treatment*

Fabrics	Afterflame s	Afterglow s	Flaming to Top or Either Side Edge	Hole Formation, Flaming, or Molten Debris
1	3	0	No	No
2	3	0	Flaming to top	No
3	0	0	No	No
4	0	0	No	No
5	0	0	No	No

TABLE 5—*Results of flame spread test for LABCYCLE treatment.*

Fabrics	Afterflame s	Afterglow s	Flaming to Top or Either Side Edge	Hole Formation, Flaming or Molten Debris
1	27	0	Flaming to top	No
2	1	0	No	No
3	0	0	No	No
4	0	0	No	No
5	0	0	No	No

Heat transfer (radiation)

The results of a heat transfer test are shown in Table 6. In the results, t_{12} is the time needed to raise the temperature to 12°C under the test specimen, and t_{24} is the time needed to raise the temperature to 24°C. prEN ISO 11611 defines minimum values for radiant heat transfer in two classes: class 1 ≥ 12s and class 2 ≥ 16s. The results after both pre-treatments are under 16 seconds but over 12 seconds. The classification of our study fabrics would therefore be class 1 according to prEN ISO 11611.

TABLE 6—*Results of radiant heat transfers test.*

Fabrics	USED		LABCYCLE	
	t_{12} s	t_{24} s	t_{12} s	t_{24} s
1	8.6	15.4	8.2	14.7
2	8.5	14.5	8.7	15.2
3	8.4	14.9	8.6	15.6
4	8.0	14.0	8.7	15.8
5	7.9	13.7	7.7	13.7

Electrical Resistance

The results of the electrical resistance test are shown in Table 7. The minimum requirement for electrical resistance in prEN ISO 11611 is $>10^5$ Ω. This level was well passed after both pre-treatments.

TABLE 7—*Results of electrical resistance test.*

Fabrics	USED Ω	LABCYCLE Ω
1	$2.09*10^{10}$	$2.43*10^{11}$
2	$1.14*10^{11}$	$3.83*10^{11}$
3	$7.09*10^{10}$	$3.66*10^{11}$
4	$1.38*10^{11}$	$2.88*10^{11}$
5	$4.54*10^{11}$	$3.62*10^{11}$

Changes in Fibers

The micrographs are shown in Fig. 4. Microscopic analysis revealed some fluff, especially on the surface of the fibers of USED treatment fabrics, and more clearly on the fibers of fabrics 3 and 4. The pictures show that the pre-treatments have affected the surface of the fibers. After both treatments, there were many cracks on the surface of the fibers of cotton fabric 1. Similar cracks were seen also on the surface of the fibers of fabric 2, which was cotton/polyester. The changes in the fibers of fabrics 3, 4, and 5 were mainly longitudinal cracks on LABCYCLE pre-treated fabrics.

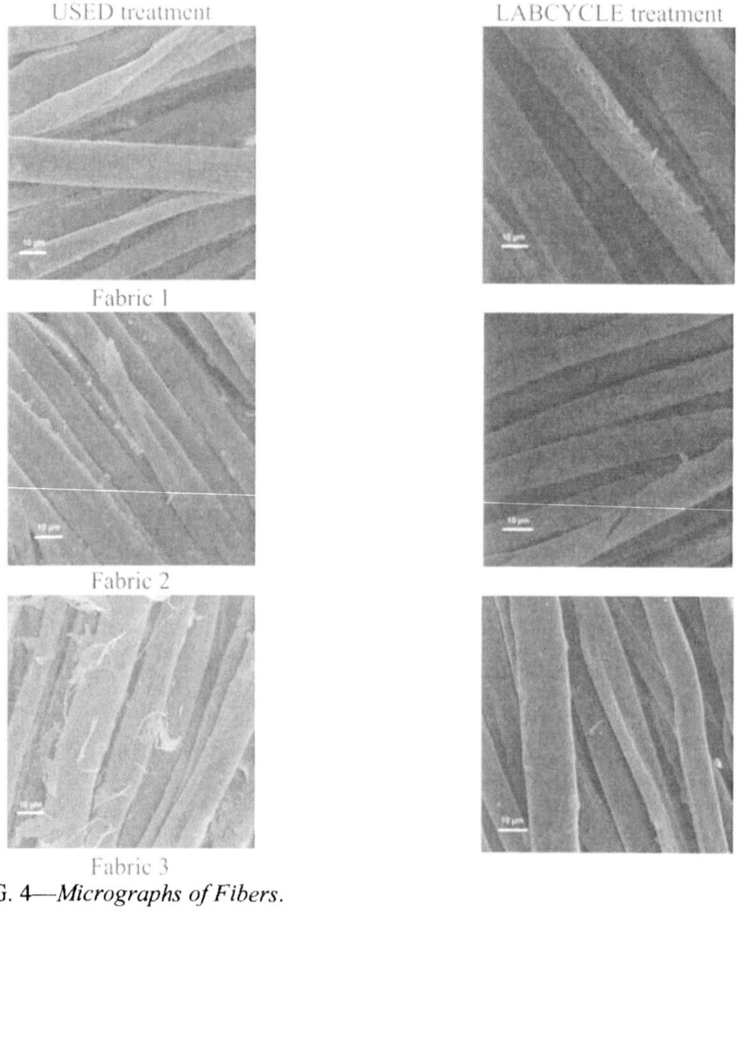

FIG. 4—*Micrographs of Fibers.*

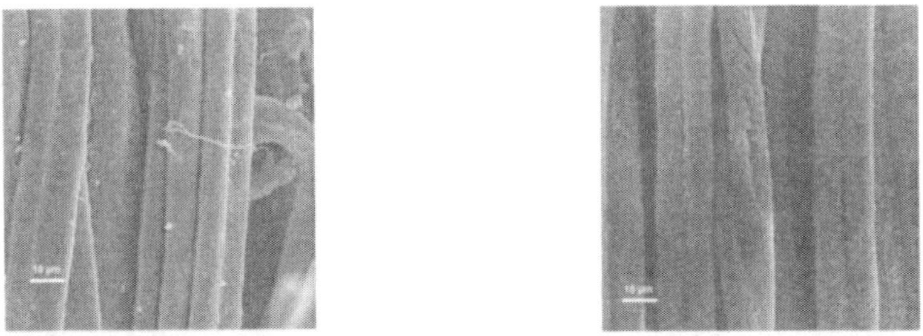

Fabric 4

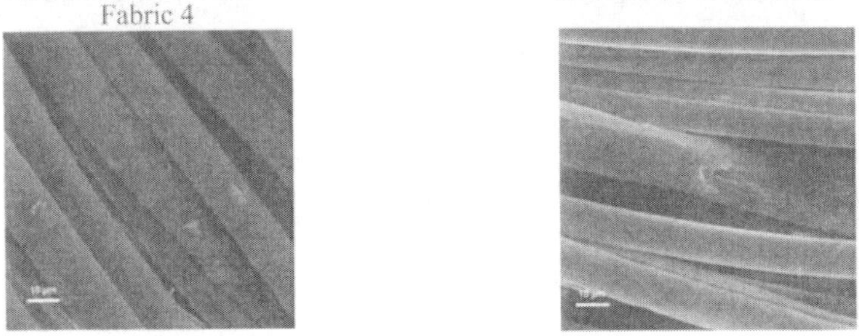

Fabric 5

FIG. 4 continued—*Micrographs of Fibers*.

Discussion

The results of the strength tests show that both pre-treatments have deteriorated the fabrics fairly equally.

The average difference between the USED and LABCYCLE pre-treated fabrics in tensile strength were ca. 100 N (16%) in the warp direction and ca. 73 N (9%) in the weft direction. For USED pre-treated fabrics 2 and 5, notably lower values were measured than for LABCYCLE pre-treated fabrics. In the warp direction, the tensile strength was much lower for USED treated samples than for LABCYCLE samples. The deviation in the results was greater for the USED specimens than for the LABCYCLE specimens. This is due to the irregular aging of the ready-made overalls during normal work at the shipyard.

The average difference in the tear strength results between USED and LABCYCLE treated fabrics was ca. 14 N (27%) in the warp direction and ca. 18 N (30%) in the weft direction. The tear strength values were systematically lower for LABCYCLE treated fabric than for USED treated fabrics. This means that LABCYCLE treatment had a greater effect on the fibers than USED treatment.

According to the results of an earlier study [12], the aging effect of continuous laboratory UVC treatment of 50 hours on welders' garment fabrics was lower than in normal use. Aging by cycling UVC/ozone and washing pre-treatments seems to better simulate the aging process of normal use than laboratory aging by UVC only. Tear strength seems to be affected by laboratory treatments more easily than tensile strength. Fabrics containing cotton lost their strength more easily in laboratory cycling pre-treatment than in normal use at the shipyard.

The limited flame spread results by surface ignition for USED and LABCYCLE treated fabrics were similar. The flame retardancy properties of fabrics 3, 4, and 5 were retained after both pre-treatments. Both pre-treatments seemed to have affected the fibers of the fabrics similarly.

No significant difference was found between USED and LABCYCLE pre-treated fabrics in the radiant heat transfer and electrical resistance tests. In the radiant heat test, the time measured for t_{12} was 7.7–8.7 sec., and for t_{24} it was 13.7–15.8 sec. The differences in this test are due to the differences in the thickness of the fabrics rather than the effect of the pre-treatment. In the electrical resistance test, the results ranged from $2.09 \cdot 10^{10}$ Ω to $4.54 \cdot 10^{11}$ Ω. The effect of the pre-treatments is negligible.

The microscopic evaluation shows that the fibers age irreversibly during exposure to UV radiation, ozone, and other agents in the welding environment. The changes are more visible in the micrographs of the cotton blend fabrics. The man-made fibers maintained their appearance to the naked eye, although the testing revealed changes in the strength of the fibers. Microscopic cross-sections might give more information about the changes in the man-made fibers.

Conclusions

Both pre-treatments deteriorated the fabrics fairly equally, with the laboratory cycling treatment having a greater effect on tear strength than on tensile strength.

The results support the conception that it is possible to simulate the aging process of normal use by laboratory pre-treatments.

Artificial aging should be included as an essential part in the analysis of the properties of welders' protective clothing to find and develop fabrics that better meet the demands of the welders' working conditions.

UVC/ozone and washing cycling pre-treatment could be a more realistic pre-treatment method in the welders' protective clothing standard than simply five laboratory washings.

Acknowledgments

The authors wish to thank the Finnish Work Environment Fund for supporting this study, and the shipyard Kvaerner Masa-Yards, Helsinki and clothing factory Kwintet Leijona Oy for their cooperation.

References

[1] Roy, C.R., "Ultraviolet Radiation: Sources, Biological Interaction and Personal Protection," *Non-ionizing Radiations, Physical Characteristics, Biological Effects and Health Hazard Assessment*, International Radiation Protection Association, Repacholi, M.H., Ed., 1988, pp. 49–70.

[2] Hietanen, M., Engström, B., Kauhaniemi, P., Mäkinen, H., "Hazards from Ultraviolet Radiation and Ozone and Their Prevention in Aluminium Welding," Finnish Institute of Occupational Health, Helsinki, 1997 (in Finnish).

[3] Welding Hazards, http://www.afscme.org/health/faq-weld.htm. Viewed on June 26, 2003.

[4] Sliney, D.H., Benton, R.E., Cole, H.M., Epstein, S.G., Morin, C.J.,

"Transmission of Potentially Hazardous Actinic Ultraviolet Radiation Through Fabrics," *Applied Industrial Hygiene 2*, 1987, pp. 36–44.

[5] Spear, J., "Evaluating Potentially Harmful Fume And Gas Exposures of Welding", ECS Risk Control Library, http://www.ecsclaims.com/hot/evaluating_potentially_harmful.htm. Viewed on June 23, 2003.

[6] Okuno, T., Ojima, J., Saito, H., "Ultraviolet Radiation Emitted by CO_2 Arc Welding", *The Annals of Occupational Hygiene*, Vol. 45, No. 7, 2001, pp. 597–601.

[7] Tenkate, T.D., and Collins, M.J., "Personal Ultraviolet Radiation Exposure of Workers in a Welding Environment," *American Industrial Hygiene Association Journal 58*, 1997, pp. 33–38.

[8] ISO 1431-1: 1982. "Rubber, Vulcanized. Resistance to Ozone Cracking. Part 1: Static Strain Test," International Organization for Standardization, Case postale 56, CH-1211 Genève, Switzerland, pp. 7.

[9] European Council Directive, 89/686/ EEC, *Official Journal of the European Communities*, December 30, 1989.

[10] EN 470:1994, Part 1. "Protective Clothing for Use in Welding and Allied Processes – General Requirements," CEN, European Committee for Standardization, rue de stassart, 36-1050 Brussels, pp. 8.

[11] Schlüter, R., Jung, K., et al., "Untersuchungen zur Schutzwirkung von Schweißerschutzanzügen," BIA-Report 4/86, Berufsgenossenschaftliches Institut Arbeitssicherheit - BIA, Lindenstrasse 80, 25205 Sank Augustin 2, Germany.

[12] Mäkinen, H., Raivo, S., et al., "Developing Requirement and Test Methods for Welders' Protective Clothing from the User's Viewpoint," *Environmental Ergonomics VIII*, International Series of Environmental Ergonomics, Vol. 1, Hodgdon, J.A., Heaney, J.H., Buono, M.J., Eds., 1998, pp. 277–280.

[13] Toyo, Protection, http://www.ehime-iinet.or.jp/ehime_e/corp/toyo/ronbun/ronbun.htm. Viewed on June 16, 2003.

[14] EN ISO 6330:2000. "Textiles – Domestic washing and drying procedures for Textile Testing," CEN, European Committee for Standardization, rue de stassart, 36-1050 Brussels, pp. 35.

[15] prEN ISO 11611:2002, October, "Protective Clothing for Use in Welding and Allied Processes," CEN, European Committee for Standardization, rue de stassart, 36-1050 Brussels, pp. 15.

Journal of ASTM International, June 2004, Vol. 1, No. 6
Paper ID JAI12119
Available online at www.astm.org

Peter A. Thorpe[1] *and David A. Torvi*[1]

Development of Non-Destructive Test Methods for Assessing Effects of Thermal Exposures on Fire Fighters' Turnout Gear

ABSTRACT: The clothing that serves to protect a fire fighter in the line of duty experiences a wide range of exposures. Specifications for turnout gear focus primarily on pre-use performance. There is no standard for the performance of in-use fire fighter protective clothing. This paper outlines a research program designed to develop non-destructive test methods to provide users with an indication of how their in-use gear would conform to standards for new gear. Three non-destructive test methods (Raman luminescence, digital image analysis and colorimetry) are developed and correlated with established destructive tests to indicate useful service life. The paper concludes that among the methods examined in this study, digital image analysis offers the most promise for providing an economical and effective technique to assess the condition of in-use garments.

KEYWORDS: protective clothing, heat transfer, non-destructive testing, turnout gear, fire fighter, degradation, digital image analysis

Introduction

The environment that a fire fighter experiences in the normal course of duty ranges from benign to very dangerous. The protective clothing that fire fighters wear (known as turnout gear) must meet a variety of performance criteria: it must protect from the extreme heat of fires, it must protect from the cold of winter, and it must be light enough that the sheer weight of the gear does not cause undue stress on their bodies. To this end, one design has become the standard for fire fighter turnout gear. This design includes a heat resistant outer shell, typically made of some synthetic fiber; a moisture barrier, typically made of a water resistant breathable material; and a thermal barrier, typically made of wool felt with some aramid fiber woven into it.

A number of standards organizations have tests for the performance of thermal protective clothing when it is new, however there is currently no standard for the continuing performance of this gear. The National Fire Protection Association has published the NFPA Standard on Protective Ensemble for Structural Fire Fighting (NFPA 1971), which specifies construction and minimum performance requirements for structural fire fighting gear. The standard references a number of ASTM fabric tests for use in the specification of new gear elements. As the performance of this clothing can deteriorate over its useful lifetime, there is a need to develop methods to determine when fire departments should retire their gear [1].

Besides safety issues, cost is another motivation for this research. The cost of turnout gear represents a very significant economic investment for fire departments. For this reason, non-destructive testing methods are desirable. The goal is to develop tests that could quantitatively determine the suitability for service of a set of in-use turnout gear without destroying the gear

Manuscript received 13 November 2003; accepted for publication 22 April 2004; published June 2004. Presented at ASTM symposium on Performance of Protective Clothing: Global Needs and Emerging Markets: 8th Symposium on 13 January 2004 in Tampa, FL.

[1] M.Sc. Candidate and Associate Professor, respectively, Department of Mechanical Engineering, University of Saskatchewan, Saskatoon, SK S7N 5A9.

during testing. For these reasons, the focus of this research is the development of new non-destructive test methods by which gear that is currently in use can be tested to determine whether it can still provide adequate protection for the fire fighter.

This paper will briefly examine current practices in the assessment of degradation of fire fighter's turnout gear, then it will move toward a discussion of research to develop non-destructive test methods for this assessment. It will describe two full-scale fire tests which were used to determine the temperatures and heat fluxes to which turnout gear might be subjected during service. The paper also will describe the fabric specimens that were tested and the artificial aging process to which they were subjected. Some destructive tests were conducted on samples of materials used for the three layers in a set of turnout gear. The focus of this paper is on tensile testing of outer shell material. Non-destructive test methods used in this research will then be described.

Previous and Current Practices

A review of previous research in the area of degradation and retirement of turnout gear can be found in [1]. One example of this research is a study of the properties of twenty sets of retired turnout gear [2]. It was observed that the gear performed well in most of the thermal tests but failed in some mechanical tests. However, this work did not examine the exposure history of the gear; rather all the gear was studied without any detailed knowledge of its history.

Previous research, along with experience in the field, has led to the development of several indicators that are currently used by fire departments to determine whether to replace turnout gear. Outlined in NFPA Standard on Selection, Care, and Maintenance of Structural Fire Fighting Protective Ensembles (NFPA 1851), the inspection protocol requires routine inspection to look for soiling, contamination, rips, tears, cuts, and thermal damage such as charring, burn holes, and melting. Every twelve months NFPA 1851 requires a more advanced inspection, which includes an examination of garment damage, as evidenced by rips, cuts, tears, discoloration, and thermal damage. Decisions are primarily based on visual inspections. Observations that indicate a garment should be retired include discoloration and brittleness, but this leaves much to the discretion of the examiner. Additionally, some degradation may occur on the microscopic level, so these visual indicators may not be capable of identifying all pieces of turnout gear that should be retired.

Much of the testing of the fire protection offered by these ensembles is done using a heat flux of 80 kW/m^2, a heat flux chosen because it was thought to represent a typical heat flux for flash fire conditions [3]. When garments are exposed to this level of heat flux, they degrade very quickly and visually appear to be damaged. Of interest is the influence of lower heat fluxes on the garments, low enough that visual signs of degradation may not be as obvious. One test that is conducted at a lower heat flux is the ASTM Standard Test Method for Radiant Protective Performance of Flame Resistant Clothing Materials (ASTM F 1939), a test that uses both 21 kW/m^2 and 80 kW/m^2 heat fluxes.

In his depiction of fire fighting conditions, Veghte [4] defines heat flux exposures up to 2.1 kW/m^2 as "routine", exposures from 2.1 to 21 kW/m^2 (with temperatures below 300°C) as "ordinary", and exposures above 12.6 kW/m^2 (with temperatures higher than 300°C) as "emergency". The focus of this research is to examine the effect of relatively low to moderate heat flux exposures on the performance of turnout gear. While the effects of these exposures on all three layers were investigated, the emphasis of this paper is the effect of these heat flux exposures on outer shell materials.

Materials Tested

A fire fighter's protective ensemble is typically constructed of three layers of fabric: the outer shell, the moisture barrier, and the thermal liner. The outer shell is designed to provide protection from the elements without breaking down. Outer shells are usually made of synthetic fibers that incorporate high strength and excellent thermal resistance. The moisture barrier is designed to protect the fire fighter from external fluid exposures. As technology has improved, these thermal barriers are now designed to allow water vapor to be removed from the body, reducing the chance of injury due to steam burns on the skin of the fire fighter. The thermal liner can be removable or permanent and is designed to provide an insulating layer to protect the fire fighter from heat or cold.

The outer shell of the ensemble used in this testing is comprised of a blend of 60 % Kevlar®[2] (para-aramid) and 40 % Nomex® (meta-aramid). The moisture barrier is a polytetrafluoroethylene (ptfe) based membrane layer. The thermal liner consists of a Nomex® face cloth and Kevlar® batting.

Full Scale Fire Testing

In order to simulate the effects of actual fires in the laboratory, it is important to understand the conditions that a set of turnout gear may be exposed to in the field. While some data on conditions faced by fire fighters is available [4,5], additional data were gathered for the purposes of this and other fire science research conducted at the University of Saskatchewan.

In one test, temperatures were measured during a controlled burn in a small room, conducted by Saskatoon Fire and Protective Services during Fire Prevention Week in October 2002. The test is described in detail in [6]. Temperature histories at the ceiling and floor of the room were plotted to observe the rise and fall of thermocouple temperatures during an actual fire. Peak temperatures measured during this test were approximately 850°C at the ceiling and at human chest height and 750°C at the floor.

A second full scale fire test was conducted in Edmonton, Alberta in collaboration with the University of Alberta Department of Mechanical Engineering and the City of Edmonton Fire Department. An abandoned house slated for demolition was used for three controlled burns, and then completely destroyed in a fourth and final burn. The house was instrumented with heat flux gauges and thermocouples. Of note for this paper is the examination of heat flux levels outside of the living room window of the house. Figure 1 provides an example of the heat fluxes measured during the final burn by a Gardon gauge located 4 m from the base of the exterior wall and mounted 1.75 m above the ground in front of the window.

The location of the heat flux gauge was appropriate as an example of a point from which fire fighters carry out typical operations, in conditions Veghte would consider to range from ordinary to emergency. Based on these results it was decided that the specimens used for this research would be exposed to heat fluxes of 5, 10, 15, 20, and 30 kW/m^2.

[2] Kevlar® and Nomex® are registered trademarks of E.I. du Pont de Nemours and Company, Wilmington, DE.

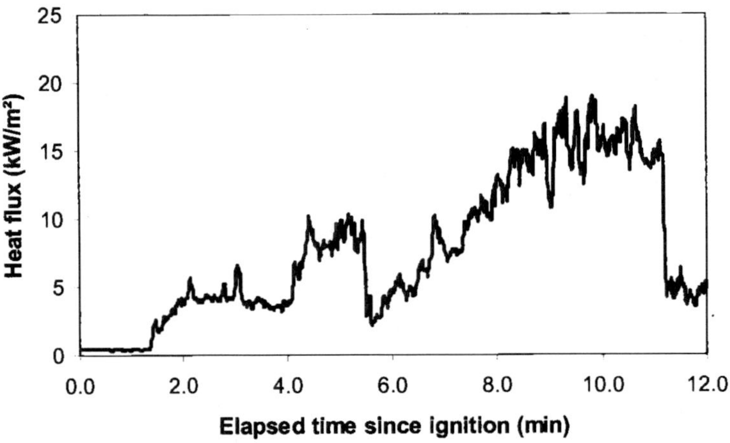

FIG. 1—*Heat flux data from Edmonton house burn.*

Laboratory Exposures of Ensemble Specimens

Fabric specimens were exposed to relatively low heat fluxes for a variety of times using the radiant panel apparatus described in ASTM F 1939.

The ASTM F 1939 stand consists of a bank of five quartz tubes that are controlled by a variable voltage supply. A modification was made to this apparatus to include a water cooled shutter. The three layer fabric ensemble was conditioned in a temperature and humidity controlled room for at least 8 h, then placed in the apparatus. A series of tests were run for each of the exposure levels that were selected. Table 1 summarizes the heat exposures to which the ensemble samples were subjected. Three specimens were subjected to 10 kW/m^2 for 600 s, and three specimens were subjected to 20 kW/m^2 for 90 s. Only a single specimen was subjected to each of the other exposures listed in Table 1. Exposure times were chosen to provide a range of degradation. For the lowest exposure level, 5 kW/m^2, testing was conducted at 10 min intervals and eventually stopped because the visible degradation was unchanged from 10 min to 60 min. For the highest exposure level, 30 kW/m^2, testing culminated at 120 s because the fabric was so charred and brittle that it could be deemed to be totally degraded.

TABLE 1—*Heat flux levels and durations used to expose fabric specimens.*

Heat Flux (kW/m^2)	Exposure Time (s)							
5	600	1200	1800	2400	3000	3600		
10	120	240	480	600*	720	900		
15	60	90	120	150	240	360	600	
20	30	45	60	75	90*	105	120	180
30	30	45	60	120				

*Three samples of each of these specimens were exposed to this heat flux and duration.

Testing Protocol

To examine the performance of the fabric specimens before and after the laboratory exposures, test requirements were selected from requirements used for new garments. NFPA 1971 provides a test matrix that gives a series of tests and minimum performance criteria for turnout gear. Most of these tests are destructive in that they require cutting of the samples and typically further destruction in the course of the testing. In addition to these destructive tests, non-destructive tests were also selected. It was decided to focus on the visual properties of the outer shell for the non-destructive testing because this was the most accessible point on the turnout gear. Developing an enhanced visual testing method also serves to complement NFPA 1851 and to help reduce inconsistencies due to the subjective nature of some of the methods of inspection.

Non-destructive test results were also compared to destructive test results in order to provide meaningful information about the condition of a particular fabric specimen. The destructive tests were chosen from those specified in NFPA 1971. While Table 2 lists the tests employed in this research, the focus of this paper is to compare the results of the non-destructive tests with results from tensile tests of the outer shell. Descriptions of the other destructive test methods and the results that were obtained can be found in [7].

TABLE 2—*Tests used to evaluate fabric specimens after exposures.*

Destructive Tests	Non-Destructive Tests
ASTM D 5034: Standard Test Method for Breaking Force and Elongation of Textile Fabrics (Grab Test)	Raman Spectroscopy
ASTM D 5733: Standard Test Method for Tearing Strength of Nonwoven Fabrics by the Trapezoid Procedure	Digital Image Analysis
ASTM F 903: Standard Test Method for Resistance of Protective Clothing Materials to Penetration by Liquids*	Colorimetry
ASTM F 1060: Standard Test Method for Thermal Protective Performance of Materials for Protective Clothing for Hot Surface Contact	

*While a test for liquid penetration is not necessarily destructive, the standard requires the use of a standardized outer shell and thermal liner, requiring the tester to cut the moisture barrier out of an in-use ensemble.

Tensile Testing (Grab Test)

The first test used to assess the performance of the turnout gear was the tensile test. The NFPA 1971 standard for fire fighter's turnout clothing requires that the outer shell material must achieve at least 623 N (140 lbf) of breaking force when tested in accordance with ASTM Standard Test Method for Breaking Force and Elongation of Textile Fabrics (Grab Test) (ASTM D 5034). This standard tests 101.6 mm (4 in.) × 152.4 mm (6 in.) specimens, and, using a constant crosshead speed of 20 mm/min, applies tension to the fabric until it fails. All of the outer shell specimens that had been subjected to radiant heat exposure were tested. The specimens with little visible signs of degradation broke at a relatively high load, while specimens with increased visible signs of degradation broke at relatively lower loads. Curves generated by this test are plotted in Figs. 2–4 for fabric specimens that were exposed to heat fluxes of 5, 10, and 20 kW/m^2. The NFPA 1971 minimum requirement for new turnout gear, 623 N, is superimposed on the figures to show the point at which the fabric will no longer meet this particular standard.

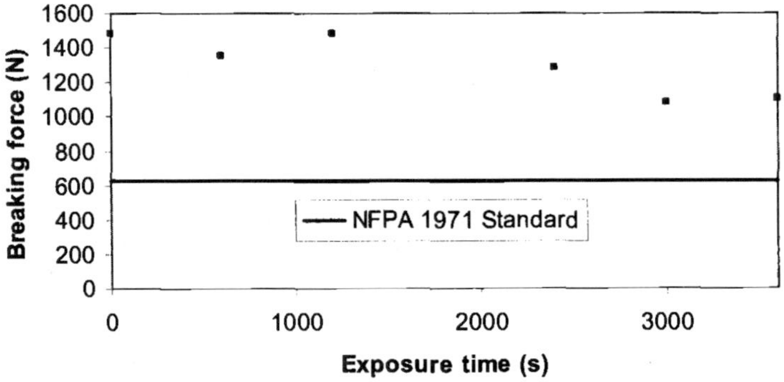

FIG. 2—*Tensile testing after 5 kW/m² radiant exposure.*

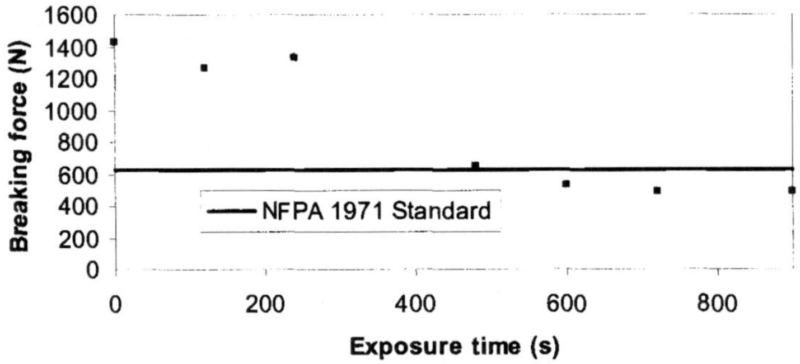

FIG. 3—*Tensile testing after 10 kW/m² radiant exposure.*

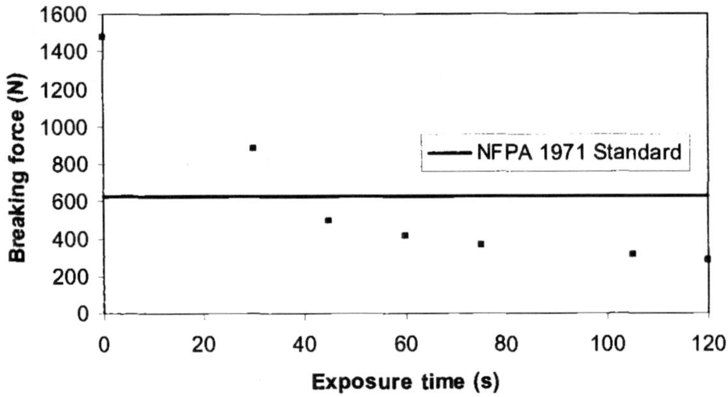

FIG. 4—*Tensile testing after 20 kW/m² radiant exposure.*

Non-Destructive Testing

The financial demands on a fire department are vast, so any way to reduce the total cost of turnout gear without compromising safety would be a welcome improvement. Currently, testing of used turnout gear can be done either by conducting destructive testing or by using primarily visual methods to assess damage. As directed by NFPA 1851, users are to inspect their gear routinely and to subject the gear to a more detailed inspection once every 12 months. The standard suggests that visual indications, such as charring and brittleness, should be used to identify degradation, but as Slater [8] suggests, this may not be soon enough for maximum safety of the fire fighter. Destructive testing can be used effectively if a number of garments have been exposed to similar conditions over their lifetimes and will behave in a similar fashion when tested, but this is an impractical approach since it assumes a great deal about the exposures and still demands that at least one garment is destroyed, making it prohibitive for all but the largest departments. For these reasons, this research has turned to a preliminary study of the development of better non-destructive test methods to characterize the thermal degradation of turnout gear. As degradation may occur due to other factors such as exposure to ultraviolet radiation and laundering procedures, future work will also be necessary to evaluate the ability of non-destructive tests to characterize other forms of degradation of turnout gear.

Raman Luminescence Technique

Raman spectroscopy is a method frequently used to examine the excitation of a material in the visual light spectrum. This method is used for surface measurements of spectra, as it does not require a translucent material. For the fabric specimens in this study it was determined that Raman spectroscopy would not be particularly useful since the fabrics responded to two different incident lasers with a very high level of fluorescence. This created a poor signal to noise ratio that reduced the effectiveness of the method.

An alternative method was devised using the same Raman microscope (a Renishaw Raman Microscope System 2000, with 514 nm excitation frequency) to quantify the luminescence of the fabric. In this method an incident laser of 514 nm illuminated the fabric, and a continuous grating scan of the emitted light was conducted from 520–1000 nm. This produced a histogram of emitted and reflected photons in the visual spectrum. It was hypothesized that the peak of this curve could be used as a diagnostic test to determine the level of degradation. Initial results were promising, as the peak wavelength appeared to shift slightly with increasing degradation. An example of the data produced by this method is given in Fig. 5. The peak is seen to shift from lower wavelength to higher wavelength as the exposure levels increase. Further examination of the repeatability of this method determined that the location of these peaks and the shape of the curves were very inconsistent. For the same unexposed specimen, the location of the peak wavelength might shift 40 nm, and the curve might have one or two large peaks (see Fig. 6). Due to this inconsistency, the method was determined to be ineffective in precisely identifying peak wavelengths for the fabric tested, and therefore may be inappropriate as a diagnostic test of degradation.

It should also be noted that the Raman microscope measures changes that occur over a very small area. Depending on the objective lens selected, this area could be of the same order of magnitude as an individual fiber or a few yarns. This is different from the other two techniques that were evaluated in this study, which consider a much larger area on the outer shell. One disadvantage to studying a very small area is that changes to individual fibers or yarns in either unexposed fabrics (e.g., variations in dye) or exposed fabrics (e.g., variations in the maximum temperature reached) will have a larger effect on results than if a larger area was considered.

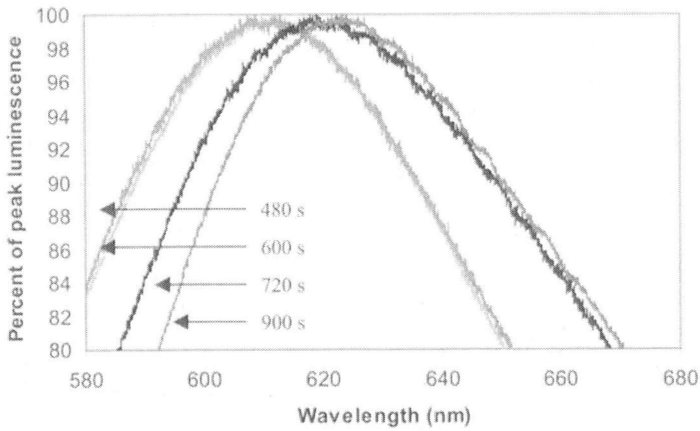

FIG. 5—*Raman luminescence spectra for outer shell fabric specimens exposed to a range of durations of 10 kW/m^2 (view focused on 580–680 nm for clarity).*

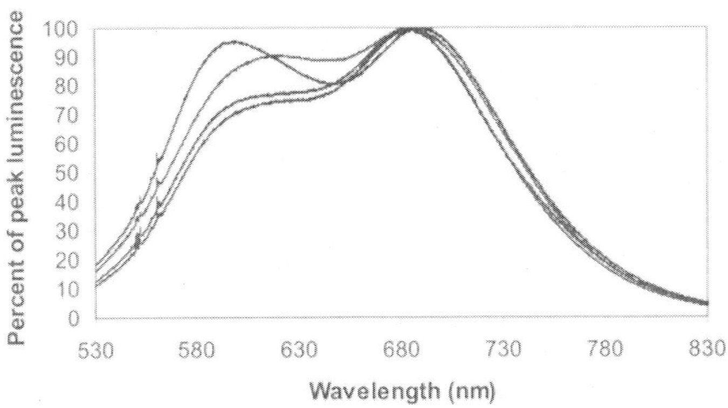

FIG. 6—*Raman luminescence method at different locations on the same unexposed outer shell fabric specimen.*

Digital Image Analysis

In this technique, exposed outer shell fabric specimens were digitally scanned, and the colors were compared and analyzed. Using a commercially available Hewlett Packard Scanjet 4400c scanner, fabric specimens were scanned with the true color setting and saved as a bitmap file. True color is based on three layers of 8-bit data, so 24 bit information for each pixel gives a possibility of 2^{24} or 16 777 216 colors. Files were saved as m × n × 3 matrices, with each m × n matrix filled with red, green, and blue intensity levels respectively. A zero level is black, and a 255 level is pure color. If all three colors are pure, then the resultant color is white. A relatively low resolution is required for this procedure, so 150 dpi × 150 dpi was used, giving file sizes in the 3 Megabyte range.

Mathworks MATLAB® software was then used to analyze the files. The program reads an image file from memory and stores it as an m × n × 3 matrix on which standard matrix operations can be performed. This particular procedure involves reading in the matrix of the fabric specimen, selecting an appropriate interrogation area, and then getting average values for the color intensity for each of the three layers. In this study an interrogation area of 100 pixels × 100 pixels was used. At a resolution of 150 dpi, this corresponds to an area of 0.67 in. (16.9 mm) by 0.67 in. (16.9 mm). The average values of each m × n matrix are calculated and returned as single values for each of the red, green, and blue layers, which represent the average values of the intensity levels of each color. The average values were then compared with color intensity averages for other heat flux exposures.

The average RGB intensity values for the different specimens are then input into a Microsoft® Excel file that compares all of the exposure levels and plots the data using a variety of criteria. In order to observe these data in an intuitive manner, the Commission Internationale de L'Eclairage (CIE) L*a*b system [9] was chosen. The CIE L*a*b system is a framework for plotting color in three spatial dimensions. The "L" value refers to lightness from 0 to 100 (from dark to light), the "a" value is the placement on a continuum from green (-100) to red (+100), and the "b" value is the placement on a continuum from blue (-100) to yellow (+100). The RGB values determined by the MATLAB® scan method were converted into CIE L*a*b values using a matrix conversion developed by the CIE [10]. The difference in color can be described using the magnitude of the vector between the two colors in this color space. The most obvious criterion to plot against is the exposure time, so there is a different graph for each heat flux level. A representative graph of the results of this analysis is shown in Fig. 7.

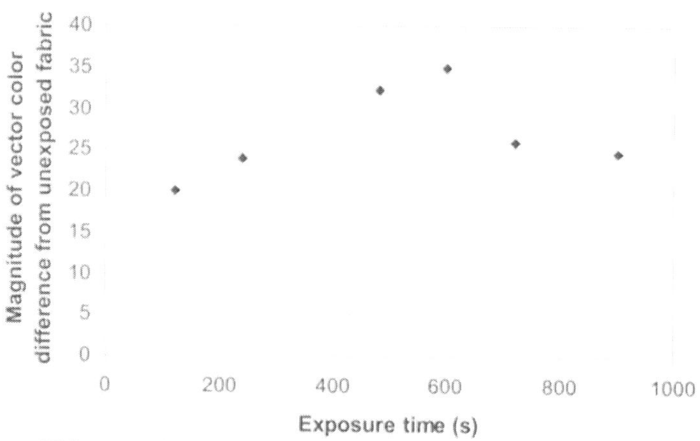

FIG. 7—*Color changes with increased heat exposure (10 kW/m^2).*

The outer shell fabric specimens in this study have been dyed light brown. This particular dye has been chosen as an appropriate compromise between visibility and reduction of the effects of soiling. As the fabric is subjected to thermal exposure, the dye is consumed and the fabric fades to a bright yellow color. After the yellowing, a charring process begins during which an increasing amount of the fabric turns black as thermal degradation occurs. The magnitude of the vector between the color coordinates in CIE L*a*b space has been chosen to represent the color difference between an unexposed specimen and the specimen in question. In Fig. 7 the rise to the highest values represent the yellowing of the fabric specimens, and the subsequent descent to smaller values represents the progression of char on the sample.

When the results from the tensile testing of the outer shell samples are plotted against the digital image analysis results, a pattern emerges that could be used to diagnose the breaking strength of a given outer shell from its color. Figure 8 gives the results from testing at exposure levels of 15 and 20 kW/m^2.

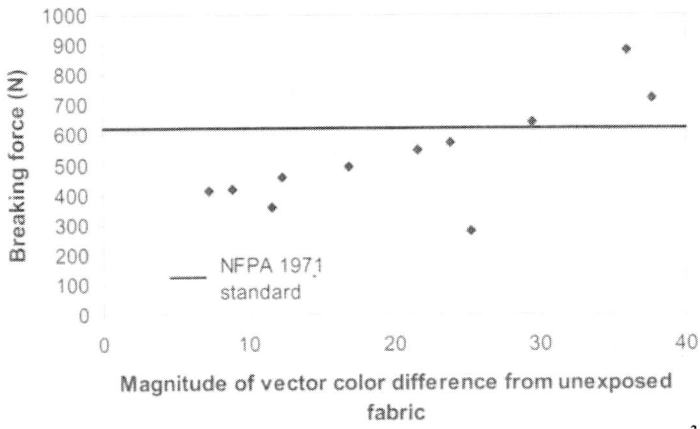

FIG. 8—*Color differences for aramid/polyaramid fabrics after 15 and 20 kW/m^2 exposure.*

In Fig. 8, the magnitude of the largest vector difference is the case where the fabric is the most yellow. The plot then descends as the fabric degrades. After a minimum magnitude at approximately 400 N of breaking force, the magnitude increases again. This is due to changes in the appearance of the char that forms on the surface of the specimen.

This technique obviously has potential, as the breaking force of a sample from in use turnout gear could be predicted from similar curves which relate changes in the magnitude of the vector color difference to the breaking load. This technique would require only a simple computer scan and some computational analysis. The sensitivity of this analysis to soil on the clothing and the variability of different scanners has not been assessed to date, but technique's promise is noted.

Colorimeter Analysis

A colorimeter offers another method of analyzing the color changes in the fabric resulting from heat exposure. For this study, a Hunterlab™ Color Analyzer was used with a 10° standard observer, a D65 standard white reference illuminant, and an aperture of 6.35 mm (0.25 in.). Data is expressed in CIE L*a*b format as in the case of the digital image analysis; this gives color space coordinates for each fabric selected. Using the L*a*b color measurement scheme, the difference in colors can be calculated as the magnitude of the vector between the coordinates.

In Fig. 9, the unexposed fabric has been chosen as the reference point for calculating color difference. This again demonstrates the increase in the magnitude of the vector color difference as the yellow color increases and the subsequent decrease as the char covers the fabric specimen.

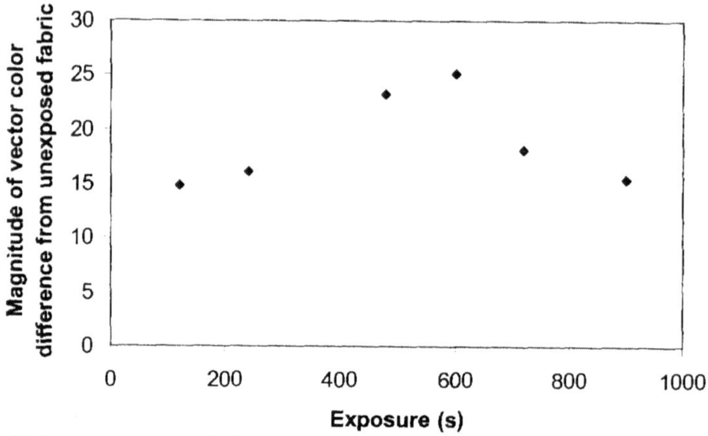

FIG. 9—*Colorimetric color difference in outer shell samples subjected to 10 kW/m² radiant exposures.*

Figure 10 shows the correlation of the results from the tensile testing and the colorimeter. The same general trend was observed in this correlation as was observed in the correlation between the tensile test results and the measurements made using digital image analysis (Fig. 8). The curve descends initially from high breaking force at low color difference to a maximum color difference at the point where the fabric is most yellow in color. The curve then shows a trend to reduced color difference as the char increases and the required breaking force is reduced. These results indicate that garments could be tested with a colorimeter, and a similar curve could be used to predict the performance of the garment in tensile testing.

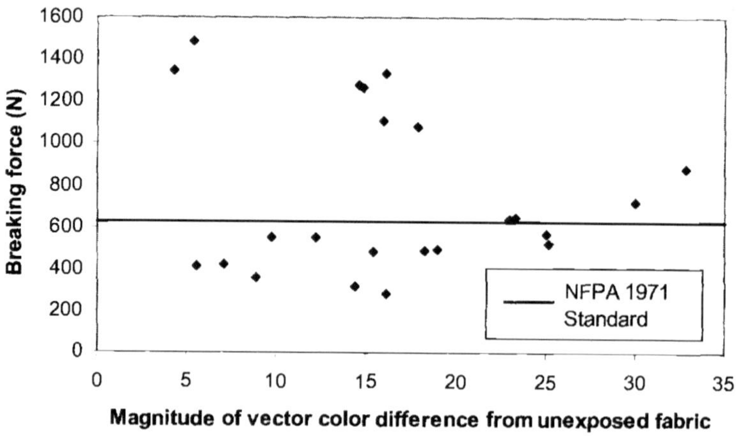

FIG. 10—*Color difference as an indicator of breaking load for outer shell fabric specimens after 10 and 15 kW/m² exposures.*

Discussion

The goal of developing a non-destructive test method for use in the fire service is to produce a test that could be easily conducted on all of the garments used by an individual department. One suggestion is to produce a test that could be brought to the fire hall by an independent contractor, reducing the capital cost to the fire department. Another option would be to develop a test that is economical enough that the equipment and any additional materials that are required could be purchased by an individual department. Of the three non-destructive test methods considered in this paper, the digital image analysis is the most economical, as it requires a low initial set-up cost to develop a program that could be run by members of the fire service. The colorimetric analysis would be less economical, as handheld colorimeter devices cost a few thousand dollars. This option would probably be most appropriate to the independent contractor who would own the colorimeter and any necessary computer software. The Raman luminescence option is the least economically feasible of the three methods. While Raman microscopes can be outfitted with remote sensors that could be used for this type of testing, a package could cost more than $50 000. Unless there is a laboratory that owns a Raman microscope in the area of a fire department, this method is probably not suited for this application.

The question of the repeatability of these test methods is not easily resolved. The Raman luminescence method showed a definite lack of repeatability for the case of unexposed outer shell fabric (Fig. 6). Testing at higher levels of degradation showed more repeatable results, but still not enough difference in the spectral peaks to be used as a diagnostic test. One suggestion is that the initial color of the fabric specimen may play a large role in determining whether the method will work. Brown is a difficult color to classify, so fabrics with more primary colors may be better suited to this analysis [10]. Further testing into green and blue Nomex® fabrics has shown that the initial color of the fabric influences the repeatability of the method, as three tests each of unexposed green and blue Nomex® specimens gave peaks within 5 nm of each other.

For the digital image analysis technique, the testing showed very repeatable results within the same sample. Questions arise, however, in the correlation with the tensile testing. There appeared to be approximately 10 % scatter in the tensile test results of similarly exposed fabrics, a variation that limits the method's effectiveness as a diagnostic test. One solution is to correlate the digital image analysis technique with some other destructive test. Work is ongoing to examine this possibility.

The colorimetric technique showed that within the same specimen, results were consistent within 1 %, but results from different specimens with the same exposure had a scatter of up to 10%. This scatter may also be resolved by using turnout gear that is a different color. The scatter of the tensile testing results likely plays a significant role in this analysis; the solution to this problem is again to look for more appropriate destructive test methods for correlation.

As discussed earlier, each of the three techniques included in this study examines a different size of area on the outer shell. The Raman microscope measures changes to an area ranging from the size of an individual fiber to a few yarns. The digital image analysis and the colorimeter measure changes to much larger areas (0.67 in. (16.9 mm) × 0.67 in. (16.9 mm), and a circle of 0.25 in. (6.35 mm) diameter, respectively). These differences should be noted when comparing the results and utility of the test methods included in this study.

Conclusion

The methods developed to conduct non-destructive testing on used turnout gear had varying degrees of success. The Raman technique of luminescence was considered promising, but there was too much inconsistency in the data to suggest that it was practical for use in an actual diagnostic test method. The digital image analysis method is a promising technique with very little capital expenditure required. The colorimetry method produced data that appeared useful, though the scatter suggested more work is needed to refine the method.

This paper has presented research done to develop non-destructive test methods to assess the remaining lifetime of a set of fire fighter's turnout gear. It was observed that by correlating the results from destructive testing with results from non-destructive testing, a diagnostic non-destructive test can be developed that would indicate whether a particular piece of gear should be retired. This research has also shown that there are difficulties when using these techniques to assess brown turnout gear.

Future Work/Recommendations

One suggestion arising from this research is the use of different colors for turnout gear. Brown is a good color for reducing obvious soiling, but it creates difficulties for the measurement of color. Results from blue and green aramid outer shell materials are promising, and additional work will reveal whether these colors would influence the repeatability of these test methods. Future research will also look at the effects of heat flux exposures on additional material properties, such as the vapour permeability of moisture barriers.

While the work described in this paper has focused on better quantifying the visual degradation cues, more work should be done to assess the utility of other approaches, such as electrical properties, ultrasound and acoustic testing, and total mass loss methods. In addition, this research has only considered the effects of a single heat flux exposure. Degradation in the field can occur due to multiple exposures to heat fluxes, as well as other factors such as abrasion, laundering, and exposure to ultraviolet radiation. Further research should be conducted to assess the ability of non-destructive tests to predict the effects of these factors individually, as well as the effects of these factors occurring simultaneously.

Acknowledgments

Assistance from Chief Brian Bentley, Dave Bykowy, and other members of Saskatoon Fire and Protective Services is acknowledged. Financial support for this research has been provided by the Natural Sciences and Engineering Research Council of Canada, and both the College of Graduate Studies and Research and the Department of Mechanical Engineering at the University of Saskatchewan. Fabric specimens have been provided by Equinox Fire and Rescue of Winnipeg, MB.

Assistance from Dr. Doug Dale and Mark Ackerman of the University of Alberta's Department of Mechanical Engineering, and Dr. Betty Crown, Dr. Jose Gonzalez, Lelia Lawson, and Helena Perkins of the University of Alberta's Department of Human Ecology is acknowledged. Todd Threlfall and David Deutscher of the University of Saskatchewan's Department of Mechanical Engineering and Dr. Ramaswami Sammynaiken of the University of Saskatchewan's Saskatchewan Structural Sciences Centre also provided assistance.

References

[1] Torvi, D. A. and Hadjisophocleous, G. V., "Evaluating the Useful Lifetime of Firefighters' Protective Clothing," *Performance of Protective Clothing: Issues and Priorities for the 21st Century, ASTM STP 1386,* C. N. Nelson and N. W. Henry, Eds., American Society for Testing and Materials, West Conshohocken, PA, 2000.

[2] Vogelpohl, T. and Easter, E. P., "Post-Use Evaluation of Fire Fighters' Turnout Coats," *Fifth Scandinavian Symposium on Protective Clothing Proceedings*, 1997.

[3] Behnke, W. P., "Predicting Flash Fire Protection of Clothing from Laboratory Tests Using Second-Degree Burn to Rate Performance," *Fire and Materials,* Vol. 8, 1984, pp. 57–63.

[4] Veghte, J. H., "Fire Fighters' Protective Clothing: Design Criteria, Second Edition," Lion Apparel, Dayton, OH, 1988.

[5] Lawson, J. R., "Fire-Fighter's Protective Clothing and Thermal Environments of Structural Fire Fighting," NISTIR 5804, National Institute of Standards and Technology, Gaithersburg, VA, 1996.

[6] Thorpe, P. A., Torvi, D. A., "Towards Non-Destructive Test Methods for Fire-Fighter's Turnout Gear," *19th Canadian Congress of Applied Mechanics,* M. Epstein and L. Sudak, Eds., 2003.

[7] Thorpe, P. A., *M.Sc. Thesis,* University of Saskatchewan, Saskatoon, SK, 2004.

[8] Slater, K., "Progressive Deterioration of Textile Materials—1: Characteristics of Degradation," University of Guelph, Guelph, ON, 1985.

[9] Uroz, J., Luo, R., and Morovic, J., "Perception of Colour Differences in Large Printed Images," *Colour Image Science: Exploiting Digital Media,* L. MacDonald and M.R. Luo, Eds., John Wiley and Sons, West Sessex, UK, 2002.

[10] Fortner, B. and Meyer, T. E., *Number by Colors: A Guide to Using Color to Understand Technical Data*, Springer-Verlag Publishers, Santa Clara, CA, 1997.

Journal of ASTM International, February 2005, Vol. 2, No. 2
Paper ID JAI12117
Available online at www.astm.org

Guowen Song,[1] *R. Barker,*[2] *D. R. Grimes,*[2] *and D. Thompson*[2]

Comparison of Methods Used to Predict the Burn Injuries in Tests of Thermal Protective Fabrics

ABSTRACT: A study was conducted to compare the two methods, Henriques Burn Integral and Stoll criteria, in thermal protective performance evaluation on firefighter clothing composites exposed to various thermal hazards. The thermal hazards that the firefighter may encounter during fire fighting are low level thermal radiation and high intensity flashover fire. With the simulation of these thermal hazards in the lab, the heat flux behind exposed clothing composites are characterized with flux rise rate and peak heat flux. Comparisons were performed on the prediction differences of clothing system made using Henriques Burn integral and Stoll criteria under different conditions. The study demonstrated that in some cases less difference is predicted by the two methods, while in other cases a significant difference is observed. Several recommendations were made for the qualitative prediction of garment and fabric thermal protective performance under different situations.

KEYWORDS: thermal protective performance, radiant protective performance, Henriques Burn Integral, Stoll criteria, thermal radiation, flash fire, TPP, RPP

Introduction

Two methods are used as the basis for burn protections in evaluating garment and fabric thermal protective performance: Stoll criteria and Henriques Burn Integral (HBI). The Stoll criteria is used in ASTM F 1060, Test Method for Thermal Protective Performance for Protective Clothing for Hot Surface Contact; ASTM F1939, Standard Test Method for Radiant Protective Performance of Flame Resistant Clothing Materials (RPP); and NFPA 1971 Thermal Protective Performance (TPP) test. The Henriques Burn Integral is used in ASTM F 1930, Standard Test Method for Evaluation of Flame Resistance Clothing for Protection Against Flash Fire Simulation Using an Instrumented Manikin. The Stoll second-degree burn criteria derives from the work of Stoll and Chianta [1], which are based on the observed exposure time required to produce a second-degree burn in blackened human skin subjected to incident heat fluxes varying from 0.1 to 0.4 $cal/cm^2 \cdot s$ (4.2–16.8 kW/m^2) in intensity [2]. The Stoll curves were established by converting the total amount of energy that must be absorbed by the skin to sustain second-degree burn or cause pain for a given length of time to a temperature rise of copper calorimeter. Therefore, the Stoll curve can be used to predict second burn time and pain time by comparing with the temperature rise of copper calorimeter. Figure 1 shows the calculated Stoll curves (temperature rise) for prediction of second-degree burn time and pain time using the specific TPP sensor (see appendix). The Stoll method has advantage of simplicity; it does not need

Manuscript received 18 November 2003; accepted for publication 19 July 2004; published February 2005.
Presented at ASTM Symposium on Performance of Protective Clothing: Global Needs and Emerging Markets: 8th Symposium on 13 January 2004 in Tampa, FL.
[1] Research Associate, Center for Research on Textile Protective and Comfort, College of Textiles, North Carolina State University, Raleigh, NC 27695-8301.
[2] Professor, Special Project Director, Associate Director, respectively, College of Textiles, North Carolina State University, Raleigh, NC 27695-8301.

sophisticated numerical calculation to estimate burn injury. However, Stoll and Chianta state that in applying these data it is essential that the incident heat pulse must be rectangular, for any variation from this shape invalidates the data [1]. Additionally, use of these curves for extended exposures requires extrapolation and this method is limited to certain heat flux intensities [2].

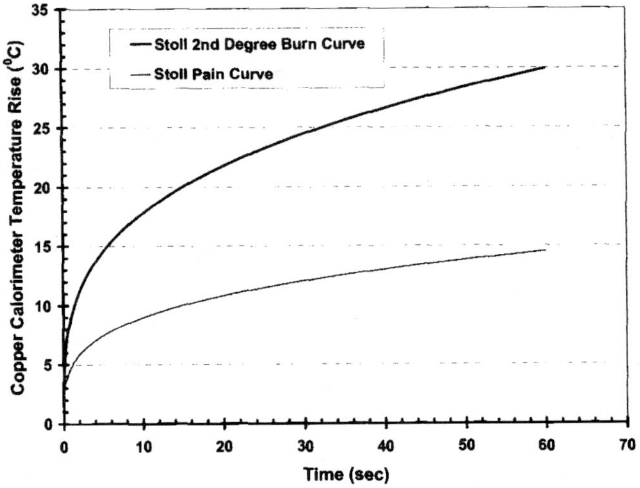

FIG. 1—*Calculated Stoll Curves for TPP Sensor.*

The Henriques Burn Integral is based on work by Henriques and Moritz [3]. Their work shows that destruction of the tissue layer located at the epidermis/dermis interface in human skin starts when the tissue temperature of the basal layer rises above 44°C. Therefore, the time that the temperature of the basal layer is above the damage temperature is critical. They found that the destruction rate can be modeled using a first order chemical reaction; that is:

$$\frac{d\Omega}{dt} = P\exp(-\frac{\Delta E}{RT}), \tag{1}$$

where
Ω = a quantitative measure of burn damage at the basal layer or at any depth in the dermis,
P = frequency factor, S^{1},
ΔE = the activation energy for skin, J/mol,
R = the universal gas constant, 8.315 J/kmol.K,
T = the absolute temperature at the basal layer or at any depth in the dermis, K,
t = total time for which T is above 44°C (317.15K).

Integration of this equation yields:

$$\Omega = \int_0^t P \exp(-\frac{\Delta E}{RT}) dt . \tag{2}$$

The integration is performed from the time when the temperature of the basal layer of the skin, T, exceeds or equals 44°C. Henriques found that if Ω is less than, or equal to, 0.5, no damage will occur at the basal layer. If Ω is between 0.5 and 1.0, first-degree burns will occur, whereas if $\Omega > 1.0$, second-degree burns will result. The damage criteria can be applied to any depth of skin provided the appropriate values of P and ΔE are used. Mathematically, a second-degree burn injury has been defined as an $\Omega > 1.0$ at the epidermis/dermis interface and a third-degree burn injury as an $\Omega > 1.0$ at the dermis/subcutaneous tissue interface. This method is used in conjunction with temperature-time data from a skin model to predict times to second and third degree burns. This method is valid for any heat flux conditions. Details about the structure of skin model and the parameters used for Henriques Burn Integral are given in the Appendix.

Data produced by decades of fire research on structural fires show that most burn injuries sustained by firefighters occurred in the low level thermal environments [4,5]. This thermal environment could be outside of the flaming envelope, postflashover or preflashover fires. The thermal intensity for this thermal environment is in the range of 0.24–0.5 $cal/cm^2 \cdot s$ (10 – 21 kW/m^2) [5,6]. Firefighters can be trapped in a flash fire or come into direct contact with flash fire for a few seconds in a rescue mission. The current firefighter turnout suits can provide 10 second protection or less to escape under most flashover conditions [7].

In this study, the two kinds of thermal hazards were simulated, and the heat flux behind the clothing composites when exposed to these hazards was characterized. The differences in times required to generate a second-degree burn made using Henriques Burn Integral and Stoll criteria in different conditions were examined.

Experimental

Materials

Two protective clothing composites, System A and System B, were selected in this study. The components of System A and System B are described in Table 1.

TABLE 1—*Protective clothing ensembles.*

Clothing System	System A	System B
Shell Fabric (g/m^2)	60% Kevlar ®/ 40% PBI, (254)	60% Kevlar ®/ 40%Nomex®, (237)
Moisture Barrier (g/m^2)	Higher THL Value Moisture Barrier (167)	
Thermal Liner (g/m^2)	One layer batting (237)	Three layer batting (271)

Sample Preconditioning

The simulation in this study was performed in dry and wet turnout systems. Dry systems were measured following sample preconditioning in a standard atmosphere (23°C ± 2°C and 60 % ± 5 % relative humidity) for 24 h. Wet systems were preconditioned using procedures that attempted to simulate the introduction of moisture by sweat produced by an active firefighter. First, the sample was conditioned for one hour in standard condition. Water was sprayed directly onto the face cloth side of the thermal liner component. Spray application was adjusted to introduce 15 % of sample weight water into the thermal liner. Then the thermal liner specimens, wetted in this manner, were laid up with outer shell and moisture barrier fabrics and sealed in a plastic bag and allowed to condition for 24 h in standard atmosphere.

Instrument and Methods

In this study two major thermal hazards were simulated: low level thermal radiation and flashover fire.

Low Level Thermal Radiation Hazard Simulation—The RPP tester was used to simulate the low level thermal radiation. The RPP tester was set up according to ASTM F 1939, (Fig. 2). The heat source was provided by five 500-watt quartz tubes. The temperature rise versus time and heat flux was measured using a copper calorimeter located behind the sample fabrics (ensembles) at a distance of 2.54 cm (1 in.) to the surface of the quartz tubes. The heat source was calibrated according to the procedure called for in ASTM F 1939. Tests were conducted on System B composite in dry and wet conditions with heat exposure levels of 0.5 cal/cm^2·s (21 kW/m^2) and 0.25 cal/cm^2·s (10.5 kW/m^2). Three sample replicates were performed for each condition.

(a) (b)

FIG. 2—*RPP tester: (a) Heat source, (b) Sample mounting and Copper Calorimeter.*

Flash Fire Simulation—A TPP tester was employed to simulate the flash fire thermal hazard. The TPP tester consists of heat source, copper calorimeter, and data acquisition system. The heat source is a combined heat flux provided by two propane fuel-fed Meker burners and a bank of quartz tubes (Fig. 3). The temperature rise and heat flux were measured using the copper calorimeter. A pneumatic shuttering mechanism activated by digital timer was used to control the exposure time. Considering that in a real firefighting situation a firefighter could get wet (sweating) and the fact that an air gap may exist between the garment and human body, this testing was conducted on System A composite with different conditions (dry and wet) and sample configuration (without spacer and with spacer). These tests were performed in accordance with the procedures established in Section 6-10 of NFPA 1971, Standard on Protective Ensembles for Structural Fire Fighting. These tests involved exposure of the composites with dry and wet preconditions to a 2.0 cal/cm^2·s (83.7 kW/m^2) heat flux with a measurement of burn threshold times using both the Henriques Burn Integral method and Stoll criterion. Two configurations (with spacer and without spacer) were also involved in this study. The nominal air gap was 6.3 mm with a spacer. Three sample replicates were performed for each condition.

FIG. 3—*TPP tester.*

Results and Discussion

Low Radiation Thermal Conditions

Two low levels of radiant heat exposure were employed in low-level thermal conditions: 0.25 cal/cm^2.s (10.5 kW/m^2) and 0.5 cal/cm^2.s (21 kW/m^2). The typical heat flux profiles behind the protective fabric composite exposed to these low level thermal exposures are illustrated in Fig. 5. The peak heat flux under the protective fabrics was found in the range of 2000–4000 W/m^2 (0.05–0.1 cal/cm^2·s) in about 40–60 s. The overall rate of heat flux rise observed is about 40–80 W/m^2/s. The rate of heat flux rise is relatively rapid in the initial 5–10 s of the test, then slows until equilibrium (steady state) is achieved. The charts suggest that heat transfer in the turnout clothing system reaches the equilibrium within the time of 50–60 s for this low level exposure. A second-degree burn was predicted for both cases.

Table 2 compares burn predictions made using both Stoll and Henriques Burn Integral methods for exposure of 0.5 and 0.25 cal/cm^2·s in dry and wet conditions. For 0.5 cal/cm^2·s level, two exposure times (40 s and 60 s) were used. In the case of the 60-s exposure time, both methods predict second-degree burns in dry and wet conditions. About a 10-s difference in second-degree burn time was identified in both conditions. For the 40-s exposure case, however, the Stoll criterion predicts no second-degree burn while Henriques Burn Integral predicts a second-degree burn at 57.8 s for dry and 50.7 s for wet. It should be noted that for the 40-s exposure, the predicted burn occurred after the exposure (flux-off stage). This predicted burn time in the cooling phase indicates that the thermal stored energy in turnout clothing system contributes to burn injuries. For 0.25 cal/cm^2·s level with exposure time of 180 s, a large difference was found in the time of second-degree burn predictions. Because the heat source level is low, the prediction time as shown in Table 2 occurs in a relatively longer time (90-125 s). The difference for both dry and wet conditions is about the same.

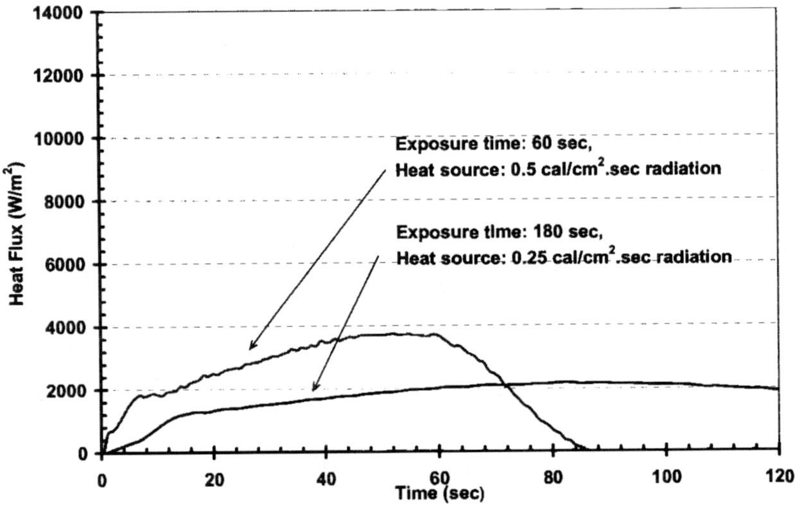

FIG. 5—*Heat flux profiles under firefighter turnout systems exposed to low radiation thermal conditions.*

TABLE 2—*Comparison of Henriques Burn Integral method and Stoll Criteria.*

Heat Source	Exposure Time* (s)	Time to Second-Degree Burn			
		Dry		Wet	
		HBI	Stoll	HBI	Stoll
0.5 cal/cm²·s	40	57.8	no burn	50.7	no burn
	60	50.3	61.9	48.8	58.2
0.25 cal/cm²·s	180	92.2**	117.3	98.4	124.9

* Data acquisition time: exposure time + 60 s.
** denotes that 3rd degree burns were also predicted.

Flashover Thermal Conditions

Figure 6 shows the heat flux profiles measured behind a turnout clothing system (dry condition) exposed to 2.0 cal/cm²·s (83.7 kW/m²) thermal conditions. A high rate of heat flux rise (380 – 580 W/m²/s) was observed for both configurations (without and with spacer). The heat flux transmission curve is characterized by high-rise rate and followed by rapid decay. The peak flux for 14-s exposure without spacer is about 6000 W/m² and occurs at about 18 s. For 18-s exposure, however, the peak heat flux could reach as high as 13000 W/m².

Tables 3 and 4 compare predictions made using Henriques Burn Integral and Stoll criterion for different test conditions and sample configurations. Different fixed exposures were performed in these test conditions and configurations. The preconditions include dry and wet (15 % moisture was added) and configuration covers without spacer and with spacer (6.4-mm air gap). Due to nonlinear nature of the Ω value calculation, larger or smaller difference in times could be observed. In contact configuration (without spacer), the predictions in both dry and wet conditions demonstrate a consistent difference in second-degree burns, except that for 10-s exposure in wet conditions no second-degree burns are predicted using Stoll criterion. The prediction differences for both dry and wet decline as the exposure time increases. This decline also suggests that the larger difference exists whenever the thermal stored energy contributes to the second-degree burn, for configuration with a spacer, as the air gap existing between the clothing system and sensor significantly increases the thermal insulation. No second-degree burns are predicted for 10- and 14-s exposure by the two methods. For 18-s exposure, Henriques Burn Integral predicts a second-degree burn while Stoll predicts no burn in both the dry and wet conditions. More than a 10-s difference in prediction is observed made using the two methods for 22-s exposure.

Figure 7 shows the Stoll curve and temperature rise measured using TPP sensor for a sample exposed to flash fire conditions for 10 s in wet condition without spacer. In this case the Henriques Burn Integral predicts a second-degree burn at 32.5 s, while Stoll criterion predicts no burn. As indicated in Fig. 7, the rise of temperature curve of the copper calorimeter behind composite sample does not reach or pass the Stoll curve, which indicates no second-degree burn predicted by Stoll curve. The Henriques Burn Integral predicts a second-degree burn at 32.5 s, which is about 22 s after exposure stops. During the cooling period (flux-off phase), the thermal stored energy in the fabric composites continue to discharge to skin and keep skin temperature rising. This was indicated in skin layer temperature change along with skin surface heat flux history as shown in Fig. 8. These curves exhibit a sustained temperature rising at the end of 10-s exposure. The difference in times required for skin to generate a second-degree burn made using the methods is attributable to two major factors. The first factor is heat flux curve behind the fabric composite. As shown in Figs. 5 and 6, these heat flux curves deviate significantly from a required "square curve." The second is the different parameters used in Henriques Burn Integral, such as the layer thickness, thermal properties in a skin model, or P and ΔE in the Integral, which can affect burn predictions.

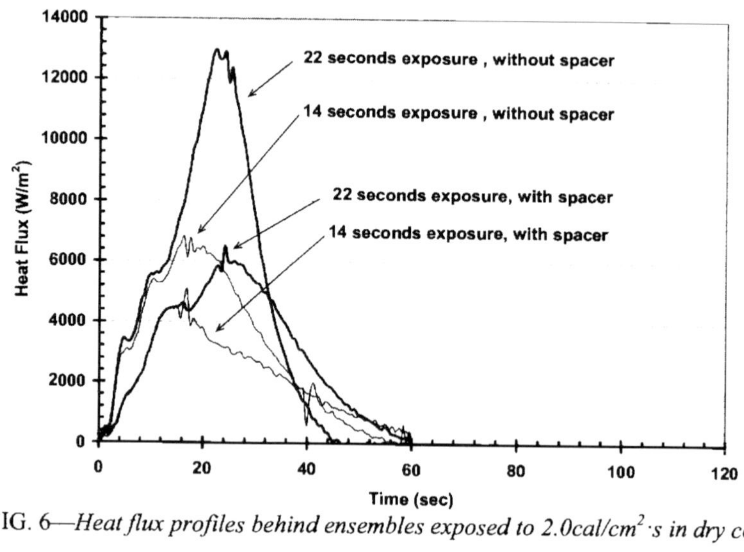

FIG. 6—*Heat flux profiles behind ensembles exposed to 2.0cal/cm*^{2}*·s in dry condition.*

TABLE 3—*Prediction comparison of HBI Method and Stoll Criterion without spacer.*

Condition	Exposure Time* (s)	Time to Second-Degree Burn (s)	
		HBI	Stoll
Dry	10	No burn	No burn
	14	23.3	35.9
	18	19.2	23.3
	22	18.9**	22.2
Wet	10	32.5	No burn
	14	17.3	22.6
	18	16.9	20.9
	22	16.7	20.4

*Data acquisition time was set at 60 s.
** Denotes that 3rd degree burn was predicted.

TABLE 4—*Prediction comparison of HBI Method and Stoll Criterion with spacer.*

Condition	Exposure Time* (s)	Time to second-Degree Burn (s)	
		HBI	Stoll
Dry	10	No burn	No burn
	14	No burn	No burn
	18	37.3	No burn
	22	28.8	39
Wet	10	No burn	No burn
	14	No burn	No burn
	18	24.2	No burn
	22	24.2	37.6

*Data acquisition time was set at 60 s.

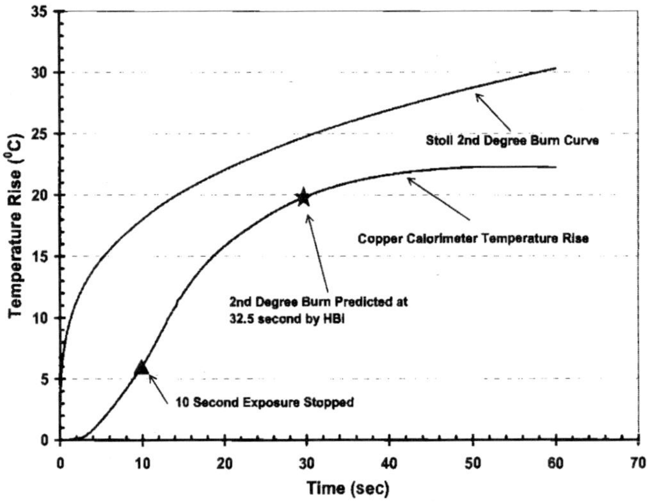

FIG. 7—*Measured temperature rise behind the composite exposed to 2.0 cal/cm²·s for 10 s in wet condition.*

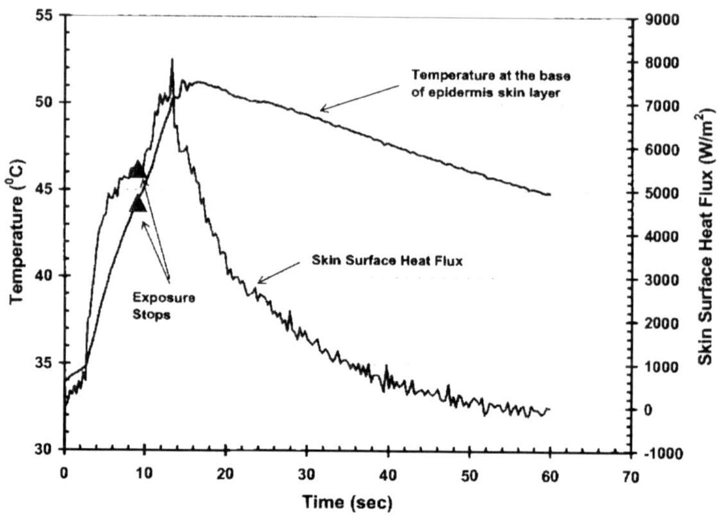

FIG. 8—*Calculated skin surface heat flux and temperature change at the base of epidermis behind the composite exposed to 2.0 cal/cm²·s (83.7 kW/m²) for 10 s.*

Conclusions

Heat flux profiles behind the turnout system exposed to different thermal conditions demonstrate different features. These heat flux profiles obtained from these exposures deviate significantly from the shape of "rectangular" or "squared."

Without spacer, the difference in second-degree burn predictions made by the two methods ranges from 3 to 5 s in flash fire conditions for relatively longer exposure time (> 18 s); for lower than this exposure a large difference was found. With spacer, more than 10-s difference was found for a 22-s exposure. In addition, on specific conditions Henriques Burn Integral predicts second-degree burn while Stoll does not. A difference of 10-25 s was predicted in low level radiation.

In flash fire conditions, the differences in second-degree burn predictions made by the two methods are pronounced in wet condition with spacer.

These data demonstrate that the difference in second-degree burn prediction made by the two methods becomes larger in low-level conditions, as well as in conditions when thermal stored energy contributes to second-degree burn. Because of the limitations of Stoll criteria that heat pulse must be rectangular and lack of data for longer exposure, this method is not recommended in burn predictions for low-level thermal hazard. It should be noted that the proper selection of a skin model and its properties, as well as factors in the Henriques integral, are crucial in the correct use of this method.

References

[1] Stoll, A. M. and Chianta, M. A., "Method and Rating System for Evaluation of Thermal Protection," *Aerospace Medication*, Vol. 40, 1969, pp. 1232–1238.

[2] Stoll, A. M. and Greene, L. C., "Relationship Between Pain and Tissue Damage Due to Thermal Radiation," *Journal of Applied Physiology*, Vol. 14, 1959, pp. 373–383.
[3] Henriques, F. C. and Moritz, A. R., "Studies of Thermal Injury: I The Conduction of Heat to and Through Skin and the Temperatures Attained Therein. A Theoretical and Experimental Investigation," *American Journal of Pathology*, Vol. 23, 1947, p. 531.
[4] Abbott, N. J. and Schulman, S., "Protection from the Fire: Nonflammable Fabrics and Coating," *Journal of Coated Fabrics*, Vol. 6, July 1976, pp. 48–64.
[5] Utech, H. P., "High Temperatures vs. Fire Equipment," *International Fire Chief*, Vol. 39, 1973, pp. 26–27.
[6] Lawson, J. R., "Fire Fighters' Protective Clothing and Thermal Environments of Structural Fire Fighting," *Performance of Protective Clothing*, Jeffrey O. Stull and Arthur D. Schwope, Eds., Vol. 6, 1996, pp. 334–352.
[7] Krasney, J. F., Rocett, J. A., and Huang, D., "Protecting Fire Fighters Exposed in Room Fires: Comparison of Results of Bench Scale Tests for Thermal Protection and Conditions During Room Flashover," *Fire Technology*, Vol. 24, February 1988, pp. 5–19.
[8] Torvi, D. A. and Dale, J. D., "A Finite Element Model of Skin Subjected to a Flash Fire," *ASME Journal of Biomechanical Engineering*, Vol. 116, 1994, pp. 250–255.
[9] Morse, H., Tickner, G., and Brown, R., "Burn Damage and Burn Depth Criteria," Aerotherm Projects 6269 and 6393, Aerotherm TN-75-26, 1975.

Appendix

Heat fluxes measured using a TPP sensor were determined from sensor temperatures using a computer program that is based on a lumped heat capacity analysis of the test sensor

$$q''(t) = \frac{mcc_l}{A}\frac{dT}{dt} + K\left(T(t) - T(0)\right)$$

where
q''= Incident heat flux (cal/cm^2·s)
m = Mass of calorimeter slug (grams)
c = Heat capacity of copper (cal/g °C)
c_l = Thickness factor as experimentally determined
A = Disk area (cm^2)
K = Heat loss coefficient as experimentally determined (cal/cm^2·s· °C)
$T(t)$= Surface temperature of disk at time t
$T(0)$= Initial or ambient temperature

A three-layer skin model was used in this study. The skin layer thermal physical properties are summarized in Table 5.

TABLE 5—*Thermal properties in the skin heat transfer model [8].*

	Human Skin	Value
Epidermis	Thermal conductivity (W/m·°C)	0.255
	Density (kg/m^3)	1200
	Specific heat (J/kg.°C)	3598
	Thickness (m)	8.0X10^{-5}
Dermis	Thermal conductivity (W/m·°C)	0.523
	Density (kg/m^3)	1200
	Specific heat (J/kg.°C)	3222
	Thickness (m)	2.0X10^{-3}
Sub-cutaneous	Thermal conductivity (W/m·°C)	0.167
	Density (kg/m^3)	1000
	Specific heat (J/kg.°C)	2760
	Thickness (m)	1.0X10^{-2}

The values of P and ΔE in Eq 2 are from Stoll and Takata [9]:

Epidermis

for $T < 50°C$ $P = 2.185 \times 10^{124}$ s^{-1}

$\Delta E/R = 93{,}534.9$ K

for $T \geq 50°C$ $P = 1.823 \times 10^{51}$ s^{-1}

$\Delta E/R = 39{,}109.8$ K

Dermis

for $T < 50°C$ $P = 4.32 \times 10^{64}$ s^{-1}

$\Delta E/R = 50{,}000$ K

for $T \geq 50°C$ $P = 9.39 \times 10^{104}$ s^{-1}

$\Delta E/R = 80{,}000$ K

Journal of ASTM International, Nov./Dec. 2004, Vol. 1, No. 10
Paper ID JAI12143
Available online at www.astm.org

Jing Zuo[1] *and Elizabeth A. McCullough*[2]

Heat Transfer Characteristics of Sports Apparel

ABSTRACT: The protective clothing and equipment required for certain sports may inhibit heat loss from the body and contribute to potentially harmful heat stress injuries. This study measured the thermal resistance (insulation) and evaporative resistance of a variety of sports ensembles worn for football, basketball, baseball, soccer, track, tennis, golf, and cycling. The thermal resistance was measured with a heated sweating manikin in an environmental chamber according to ASTM F 1291. The intrinsic insulation values (R_{cl}) ranged from a low of 0.018 $m^2 \cdot °C/W$ for running/aerobic ensemble consisting of a sports bra and compression shorts to 0.152 $m^2 \cdot °C/W$ for a football uniform. The intrinsic evaporative resistance values (R_{ecl}) ranged from 0.004–0.028 $m^2 \cdot kPa/W$ for the same ensembles, respectively. Sports uniforms that cover a large portion of the body surface or provide impact protection substantially reduced heat loss from the body. The data presented in this paper can be used in the solution of heat balance equations to predict physiological responses of athletes when wearing uniforms under different conditions.

KEYWORDS: heat stress, sports apparel, sweating manikin, athletics

Introduction

Heat stress is a common problem for athletes in sports where participants play at high intensity levels for long periods of time or under hot environmental conditions or both. The symptoms of heat stress include heat oedema, heat cramps, heat syncope, heat exhaustion, and heat stroke. Heat stroke is the most severe condition, where a person may collapse very suddenly as the sweating mechanism begins to fail and the body temperature reaches dangerously high levels. It is a major cause of death for football players, because a large amount of protective clothing is worn, and the games may be played in hot environmental conditions [1]. Heat stroke also may occur in marathon runners and soccer players when they participate in extreme environmental conditions without rest or water replenishment [2,3]. Heat syncope, anhidrotic heat exhaustion, and heat exhaustion salt depletion are less severe heat illnesses and are often diagnosed for marathon runners, soccer players, and football players. Heat cramps are commonly reported in the tennis and cycling competitions also [4,5].

League regulations and safety considerations require specialized sports apparel to be worn during competition to provide impact protection, identify teams and players, and enhance athletic performance. However, the protective garments and equipment impede dry and evaporative heat loss from the body to the environment. If the body is not able to eliminate the excess heat produced during intense physical activity, the body temperature will rise, causing potentially harmful heat stress injuries. The degree to which the protective apparel modifies the heat

Manuscript received 26 November 2003; accepted for publication 15 June 2004; published November 2004.
Presented at ASTM Symposium on Performance of Protective Clothing: Global Needs and Emerging Markets: 8th Symposium on 13–15 January 2004 in Tampa, FL.
[1] Graduate Research Assistant, Institute for Environmental Research, Kansas State University, 64 Seaton Hall, Manhattan, KS 66506.
[2] Professor and Co-Director, Institute for Environmental Research, Kansas State University, 64 Seaton Hall, Manhattan, KS 66506.

exchange between the athlete and the environment depends upon the amount of the body surface area covered by the clothing, the characteristics of the component fabrics and protective materials, and the air trapped within and between material layers. Although entire databases exist for quantifying the effects of indoor and outdoor "everyday" clothing and industrial clothing on heat exchange [6,7], there are no published data for athletic uniforms. If such data were available, they could be incorporated into so-called rational approaches to environmental safety, i.e., solving heat balance equations to predict heat storage and thermal strain [8].

Therefore, the purpose of this study was to measure the thermal resistance (insulation value) and evaporative resistance of eighteen sports ensembles using a heated, sweating manikin in an environmental chamber.

Methods

Sports Apparel Selected for Testing

Typical sports ensembles worn during official play or practice where potential heat stress problems might occur were selected for evaluation. These included eighteen ensembles for football, basketball, baseball, soccer, track, tennis, golf, and cycling. Garments for the football, basketball, and baseball ensembles were provided by a university athletic department. The other sports ensembles were purchased from specialty sports apparel retailers. The ensembles were chosen by considering applicable National Collegiate Athletic Association® (NCAA) regulations, the variety of garments worn for a particular sport, the amount of body surface area covered by the garments, and gender preferences. The garments included in each ensemble are listed in Table 1 along with the total weight of each ensemble.

Descriptions of the individual garments and their fabrics are given in Table 2. The thickness of the component fabrics was measured according to ASTM D 1777-96 Standard Test Method for Thickness of Textile Materials [9]. A 7.6 cm presser foot at 0.07 kPa of pressure was applied on the fabric with a loading time interval of 5 s. The thickness of the fabric was reported as the average of three replications. The fabric weight per unit area was determined by weighing a 5 × 5 cm square of fabric on a top loading balance and converting the weight to g/m^2.

Manikin Measurements

Apparatus—The thermal resistance (insulation) values for the sports ensembles were measured using an electrically-heated manikin in an environmental chamber. The manikin consisted of a black anodized copper skin formed to simulate the physical shape and size of a reference man (i.e., 1.8 m^2 surface area, 176 cm height). Heating wires were bonded to the inside surface of the copper skin to provide internal heating distributed so as to simulate the skin temperature distribution of a human. The skin temperature of the manikin was measured using 16 thermistors located on different parts of the body. The power cables and thermistor wires passed from the body cavity through the left side of the neck to the control and measurement equipment. The entire system was computer operated.

Procedures for Dry Manikin Tests—The thermal resistance was measured according to ASTM F 1291, Standard Test Method for Measuring the Thermal Insulation of Clothing Using a Heated Manikin, Option #1 [10]. The garments were hung on a clothing rack in an environmental chamber. The environmental conditions for the dry insulation tests were controlled as follows: ambient air temperature = 20°C, dew point temperature = 6°C (40 % relative humidity), air velocity = 0.2 m/s, and manikin surface temperature = 33.3°C.

TABLE 1—*Descriptions of sports ensembles.*

Descriptions of Sports Ensembles (Garment Code Numbers Given in Parentheses)	Clothing Area Factor (f_{cl}), unitless; Total Ensemble Weight (W_t), kg
#11: Football Uniform (Warm Weather): Helmet and chin strap (1); sleeveless cut-off T-shirt (5); shoulder pads (8); short-sleeve mesh jersey (tucked in) (3); jock strap (0); hip girdle containing hip, thigh pads, and tail bone protector (11); football pants containing knee pads and belt (12); ankle-length socks (15); athletic shoes (16).	f_{cl} = 1.25 W_t = 6.651
#12 Football Uniform (Temperate Weather): Helmet and chin strap (1); long-sleeve knit shirt (6); shoulder pads (8); short-sleeve knit jersey (tucked in) (4); gloves (9); jock strap (0); hip girdle containing hip, thigh pads, and tail bone protector (11); football pants containing knee pads and belt (12); knee-length socks (14); athletic shoes (16).	f_{cl} = 1.25 W_t = 7.022
#14 Practice Football Ensemble (with Hip Girdle): Helmet and chin strap (1); sleeveless cut-off T-shirt (5); shoulder pads (8); short-sleeve mesh jersey (cut off at the waist and hanging loosely) (2); jock strap (0); hip girdle containing hip, thigh pads, and tail bone protector (11); mesh shorts (13); ankle-length socks (15); athletic shoes (16).	f_{cl} = 1.20 W_t = 6.128
#15 Practice Football Ensemble (Shorts Only): Helmet and chin strap (1); sleeveless cut-off T-shirt (5); shoulder pads (8); short-sleeved mesh jersey (cut off at the waist and hanging loose) (2); jock strap (0); mesh knit short (13); ankle-length socks (15); athletic shoes (16).	f_{cl} = 1.20 W_t = 5.893
#21 Basketball Uniform: basketball mesh jersey (tucked in) (17); men's brief (46); basketball mesh shorts (18); footie socks (43); athletic shoes (45).	f_{cl} = 1.11 W_t = 1.743
#22 Basketball Uniform (with Shirt and Compression Shorts): sleeveless T-shirt (tucked in) (52); basketball mesh jersey (tucked in) (17); basketball compression shorts (21); basketball mesh shorts (18); ankle-length socks (35); athletic shoes (45).	f_{cl} = 1.12 W_t = 2.010
#31 Soccer Uniform: soccer short-sleeve, crew-neck jersey (tucked in) (22); men's brief (46); soccer shorts (23); soccer shin guards (25); soccer knee-length socks (24); athletic shoes (45).	f_{cl} = 1.15 W_t = 1.702
#41 Women's Aerobic/Running Ensemble: sports bra (44); above-the knee length knit tight shorts (30); footie socks (43); athletic shoes (45).	f_{cl} = 1.06 W_t = 1.253
#42 Long Distance Running Ensemble: sleeveless singlet (tucked in) (28); short shorts (29); ankle-length socks (35); athletic shoes (45).	f_{cl} = 1.09 W_t = 1.201
#43 Casual Running Ensemble: short-sleeve, crew neck T-shirt (27); loose-fitting shorts (31); ankle-length socks (35); athletic shoes (45).	f_{cl} = 1.12 W_t = 1.328
#44 Running Ensemble (Cold Weather): long sleeve zip-up shirt (zipped up) (47); long knit tights (48); ankle-length socks (35); athletic shoes (45).	f_{cl} = 1.14 W_t = 1.456
#51 Men's Tennis Ensemble: tennis cap (32); button-down sports shirt (tucked in) (33); men's briefs (46); tennis shorts (34); ankle-length socks (35); athletic shoes (45).	f_{cl} = 1.13 W_t = 1.602
#52 Women's Tennis Ensemble: tank top (tucked in) (49); women's tennis briefs (51); tennis skirt (50); footie socks (43); athletic shoes (45).	f_{cl} = 1.09 W_t = 1.356
#61 Cycling Ensemble: helmet (41); short-sleeve zip-up jersey (zipped up) (36); gloves (39); compression shorts (38); ankle-length socks (40); athletic shoes (45).	f_{cl} = 1.15 W_t = 1.650
#62 Cycling Ensemble with Shirt: helmet (41); short-sleeve T-shirt (42); short-sleeve zip-up jersey (half zipped) (36); gloves (39); loose-fitting shorts (37); ankle-length socks (40); athletic shoes (45).	f_{cl} = 1.16 W_t = 2.013
#71 Golf Ensemble: visor (67); golf shirt (tucked in) (54); golf glove (left hand) (55); men's briefs (46); long trousers (53); belt (56); ankle-length socks (35); athletic shoes (45)	f_{cl} = 1.15 W_t = 2.057
#81 Baseball Uniform (Warm Weather): baseball cap (64); short-sleeve T-shirt (tucked in) (65); short-sleeved baseball jersey (tucked in) (59); slider briefs (57); ankle-length baseball pants (rolled under up to bottom of knee) (60); fabric belt (63); knee-length socks (58); athletic shoes (45).	f_{cl} = 1.24 W_t = 2.597
#82 Baseball Uniform (Temperate Weather): baseball cap (64); long- sleeve mock turtleneck shirt (tucked in) (66); sleeveless baseball jersey (tucked in) (61); slider briefs (57); ankle-length baseball pants (62); fabric belt (63); knee-length socks (58); athletic shoes (45).	f_{cl} = 1.25 W_t = 2.628

TABLE 2—*Characteristics of garments used in the sports ensembles.* [a]

No.	Garment and Fabric Description	Fiber Content	Thickness, mm	Fabric Weight, g/m²
0	Jock strap	...	2.3	...
1	Football helmet and chin strap	...	...	...
2	Short-sleeve, v-neck mesh knitted football jersey, cut off at the waist	100 % Nylon	0.7 mesh	187.0
3	Short-sleeve, v-neck mesh knitted football jersey	100 % Nylon	0.7 mesh	187.0
4	Short-sleeve, v-neck football jersey (mesh and tricot knit fabric)	100 % Nylon	0.5 mesh 1.2 knit	179.5 mesh 430.9 knit
5	Sleeve-less knit T-shirt, cut off at the waist	58 % Polyester, 42 % Cotton	0.8	215.0
6	Long-sleeve, mock turtleneck knit shirt	91 % Polypropylene, 9 % Spandex	1.3	294.8
8	Shoulder pads	...	...	...
9	Gloves	...	3.7	...
11	Compression knit, knee length, hip girdle containing hip, thigh pads, and tail bone protector	83 % Polyester, 17 % Spandex	0.6	210.1
12	Tricot knit, knee length, football pants containing knee pads and belt	Body of pants: 90 % Nylon, 10 % Spandex	1.0	329.5
13	Mesh knit, above-the-knee length football shorts with lining	100 % Nylon	0.7 total	203.8 total
14	Knee-length terry knit football socks	...	6.6	...
15	Ankle-length terry knit socks	...	5.5	...
16	Athletic shoes for artificial turf	...	...	...
17	Basketball mesh knit jersey	100 % Nylon	1.0	238.0
18	Basketball mesh knit knee-length shorts (double layer)	100 % Nylon	2.0	502.1
21	Basketball tricot knit knee-length compression shorts	87 % Nylon, 13 % Spandex	0.9	331.1
22	Soccer short-sleeve, crew neck, knit jersey	100 % Polyester	0.7	152.8
23	Soccer tricot knit shorts, above-the-knee length, loose fitting	100 % Polyester	0.5	130.1
24	Soccer knee-length rib knit socks	90 % Stretch Polyester, 8 % Nylon and 2 % Spandex	5.0	...
25	Soccer shin guard	Cover: Polyamide Bone: Polythylene Bioplate: Polypropylene Backing: Polyester	15.9 center	...
27	Short-sleeve, crew neck, knit T-shirt	100 % Polyester	0.6	131.5
28	Sleeveless, scoop neck, mesh knit singlet (tank top)	100 % Polyester	0.4	122.8
29	Short-length, woven running shorts with mesh knit brief inside	Body: 100 % Polyester Mesh: 81 % Nylon, 19 % Spandex	0.2 body 0.5 mesh	74.1 body 114.2 mesh
30	Above-the-knee length knit tight shorts for aerobics or running	88 % Polyester, 12 % Spandex	0.8	304.4
31	Woven, above-the-knee length, loose-fitting running shorts with knit brief inside	Body: 83 % Polyester,17 % Spandex Lining: 100 % Polyester	0.4 body 0.5 lining	158.4 body 114.2 lining
32	Tennis cap	100 % Polyester	0.9	...
33	Button-down, double knit short-sleeve tennis shirt	63 % Cotton, 37 % Polyester	0.9	200.9
34	Woven above-the-knee length tennis shorts	100 % Polyester	0.3	140.5
35	Ankle-length knit socks	50 % Cotton, 35 % Acrylic, 15 %	3.8	...

		Nylon		
36	Cycling short-sleeve, zip-up, knit jersey	100 % Polyester	0.6	136.4
37	Cycling above-the-knee length, loose-fitting woven shorts with knitted brief inside	Shell: 100 % Nylon Lining: 96 % Nylon, 4 % Spandex	0.4 shell 0.8 lining	150.2 shell 185.7 lining
38	Cycling above-the-knee length, compression knit shorts	84 % Polyester, 16 % Spandex	0.7	200.6
39	Cycling gloves	...	...	...
40	Cycling ankle knit socks	44 % Polyester, 13.5 % Spandex, 17.5 % Nylon, 22.5 % Polyester, 2.5 % X-static	3.8	...
41	Cycling helmet	...	...	...
42	Short-sleeve, crew neck, knit T-shirt	100 % Cotton	0.9	216.4
43	Footie rib knit socks	79 % Cotton, 16 % Nylon, 5 % Spandex	5.4	...
44	Knit sports bra	Body: 90 % Cotton, 10 % Spandex Lining: 100 % Nylon	1.0	247.9
45	Athletic shoes	Upper: Leather, manmade Bottom: manmade materials	...	...
46	Men's rib knit brief	100 % Cotton	1.0	205.6
47	Long-sleeve zip-up knit running shirt	100 % Polyester	0.9	206.8
48	Long knit tights for running	63 % Nylon, 23 % Polyester, 14 % Lycra	1.2	222.1
49	Knit sleeveless tank top	100 % Cotton	1.3	344.0
50	Woven tennis skirt	100 % Polyester	0.2	89.2
51	Women's knit tennis briefs	90 % Cotton, 10 % Spandex	1.1	260.5
52	Sleeveless knitted T-shirt for basketball uniform	100 % Cotton	1.0	207.2
53	Long woven trousers	100 % cotton	0.7	298.5
54	Short-sleeve knitted golf shirt	100 % cotton	0.8	199.1
55	Golf glove (left hand)	Leather	0.6	...
56	Belt	Leather	4.3	...
57	Slider jersey knit briefs	90 % Cotton, 10 % Spandex	1.2	296.9
58	Knee-length baseball socks		5.5	...
59	Short-sleeve, front button, tricot knit baseball jersey	100 % Nylon	0.9	307.3
60	Long tricot knit baseball pants	100 % Polyester	0.8	293.0
61	Sleeveless front button, tricot knit baseball jersey	100 % Nylon	1.0	320.9
62	Long baseball tricot knit pants	100 % Nylon	1.1	309.3
63	Fabric woven belt		3.1	...
64	Baseball cap	100 % Cotton	1.2	...
65	Short-sleeve knit T-shirt for baseball uniform	50 % Cotton, 50 % Polyester	0.8	220.8
66	Long-sleeve mock turtleneck baseball knit shirt	100 % Cotton	0.9	240.3
67	Sun visor	100 % Cotton	4.5	...

[a] An average, double thickness of socks and gloves (that was measured in several places) is given in the table.

The air temperature was measured continuously with a matrix of four thermistors mounted on a vertical wooden stand located between the wall air supply and the manikin (about 1 m from the manikin). The dew point temperature was measured continuously with a sensor in the same location. The air velocity was measured periodically using an anemometer.

To conduct a test, the manikin was dressed in an ensemble, and all closures were secured. The manikin was hanging from its metal stand by a hook in its head. Its feet touched the floor, and its arms hung at his sides. Equilibrium was maintained for at least 1.5 h before testing and was indicated by a steady-state power reading that had not changed more than 2 % over a 30-min period. Data were collected by a computer every 30 s for a 30-min test.

The total thermal insulation value (R_t) is the total resistance to dry heat loss from the body, which includes the resistance provided by the clothing and the air layer around the clothed body. R_t is measured directly with a manikin and is calculated by:

$$R_t = (T_s - T_a) \cdot A_s / H \tag{1}$$

where R_t = total thermal insulation of the clothing plus the boundary air layer ($m^2 \cdot °C/W$), T_s = mean skin temperature (°C), T_a = ambient air temperature (°C), A_s = manikin surface area (m^2), and H = power input (W).

A clo unit is normally used for expressing clothing insulation since it is related to commonly worn ensembles. A warm business suit ensemble provides an average of approximately 1 clo of insulation for the whole body, where 1 clo of insulation is equal to 0.155 $m^2 \cdot °C/W$ [10]. When insulation values are reported in clo units, the symbol **I** is usually used instead of **R**. Therefore,

$$I_t = 6.45 \cdot (T_s - T_a) \cdot A_s / H \tag{2}$$

In this study, the total insulation value for each ensemble was reported as the average of three independent replications made on one sample set of clothing. Because R_t or I_t includes the resistance at the surface of the clothed body, it is influenced by air velocity and temperature level (as it relates to incident radiation). The factors can be dealt with easily in physiological models, which predict how much heat a person will lose or gain under a specific set of conditions. However, in some models and applications, it may be preferable to separate the resistance of the clothing from the resistance of the air layer.

Intrinsic clothing insulation (R_{cl}, I_{cl}) indicates the insulation provided by the clothing alone and does not include the insulation provided by the surface air layer. R_{cl} is defined by

$$R_{cl} = R_t - (R_a / f_{cl}) \tag{3}$$

and I_{cl} is defined by

$$I_{cl} = I_t - (I_a / f_{cl}) \tag{4}$$

where R_{cl} = intrinsic clothing insulation ($m^2 \cdot °C/W$), R_t = total thermal insulation of the clothing plus the boundary air layer ($m^2 \cdot °C/W$), R_a = resistance of the boundary air layer around the nude manikin ($m^2 \cdot °C/W$), f_{cl} = clothing area factor (unitless), I_{cl} = intrinsic clothing insulation (clo), I_t = total thermal insulation of the clothing plus the boundary air layer (clo), I_a = resistance of the boundary air layer around the nude manikin (clo).

The insulation provided by the external air layer was determined by testing the manikin nude. The f_{cl} is the ratio of the clothed body surface area to the nude body surface area. The term (R_a/f_{cl} or I_a/f_{cl}) is the resistance provided by the air layer around the clothed body. It is smaller than the air layer resistance for the nude body because the clothing increases the surface area and thus provides a greater area for heat transfer. The f_{cl} varies from 1.00 (nude) to a maximum of about 1.50 for most clothing ensembles. It can be measured with a photographic method [6] or estimated from ISO Standard 9920 [11]. In this study, f_{cl} values were determined by taking photographs of the manikin from the front and side angles, and comparing the projected area of the manikin clothed with the projected area of the manikin nude.

Procedures for Sweating Manikin Tests—The environmental conditions for the sweating manikin tests were controlled as follows: ambient air temperature = 26.7°C, dew point temperature = 12°C (40 % relative humidity), air velocity = 0.15 m/s, and manikin surface temperature = 32.8°C. For the evaporative tests, the manikin was covered with a cotton knit "skin" and sprayed with distilled water to simulate skin saturated with sweat (i.e., 100 % skin wettedness). The procedures were similar to those used for the dry tests; however, the environmental conditions were warmer. The basic equation for calculating the total resistance to evaporative heat transfer provided by the clothing is

$$R_{et} = (P_s - P_a) \cdot A_s / H \tag{5}$$

where R_{et} = resistance to evaporative heat transfer provided by the clothing and the boundary air layer (m²·kPa/W), P_s = saturated water vapor pressure at the skin surface (kPa), P_a = the water vapor pressure in the air (kPa), and H' = power input (W).

Ideally, the air temperature and the manikin temperature would be exactly the same, and all of the heat transferred would be due to moisture evaporation. However, the air temperature was about 5°C lower than the skin temperature in the sweating tests, and consequently, some dry heat transfer occurred. It is necessary to correct H for this heat transfer by the relationship

$$H' = H - \left[(T_s - T_a) \cdot A_s / R_t\right] \tag{6}$$

Therefore, the mean R_t value from the dry tests on the ensemble is needed to calculate R_{et}. The actual calculation for R_{et} is then,

$$R_{et} = (P_s - P_a) \cdot A_s / \left[H - \left((T_s - T_a) \cdot A_s / R_t\right)\right] \tag{7}$$

Data from three replications of the wet tests were averaged to determine the mean R_{et} for each ensemble (including the air layer) and for the air layer alone (tested using the wet cotton skin).

The equation for calculating the intrinsic evaporative resistance provided by the clothing alone (R_{ecl}) is analogous to that for dry resistance:

$$R_{ecl} = R_{et} - (R_{ea} / f_{cl}) \tag{8}$$

where R_{ea} = the resistance to evaporative heat transfer for a still air layer (m²·kPa/W).

Finally, the moisture permeability index (i_m) indicates the maximum evaporative heat transfer permitted by a clothing system as compared to ideal maximum from an uncovered surface (i.e., a slung psychrometer). It was defined by Woodcock [12] as

$$i_m = (R_t / R_{et}) / (R_a / R_{ea}) \tag{9}$$

The value of the ratio R_a/R_{ea} varies somewhat with temperature, but it is typically taken to have the value of 16.65 °C/kPa (i.e., the Lewis Relation). The i_m value for a clothing system (including the air layer resistance) can be calculated by the relationship

$$i_m = (R_t / R_{et}) / 16.65 \tag{10}$$

Although theoretically the i_m values could range from 0 for a system with no evaporative heat loss to 1 for a system that had little or no resistance to evaporative heat transfer, values greater than 0.50 are seldom obtainable for a motionless nude manikin in relatively still air.

Results

The resistance values for the sports ensembles are presented in Table 3. The thermal resistance (insulation) associated with the still air layer around the nude manikin (R_a) was 0.094 m²·°C /W (0.61 clo). The intrinsic clothing resistance values for the sports ensembles ranged from 0.018 m²·°C /W (0.116 clo) for a women's aerobic/running ensemble (#41) consisting of a sports bra and above-the-knee length knit tight shorts to 0.152 m$^2\cdot$°C/W (0.982 clo) for a football uniform worn in temperate weather (#12). Both football uniforms (#11, #12) provided a thermal resistance that was several times higher than the other sports ensembles.

TABLE 3—*Thermal insulation and evaporative resistance values for sports ensembles.*

Ensemble	Total Thermal Insulation R_t, m$^2\cdot$°C /W I_t, clo	Intrinsic Clothing Insulation R_{cl}, m$^2\cdot$°C /W I_{cl}, clo	Total Evaporative Resistance R_{et}, m$^2\cdot$kPa/W	Evaporative Resistance of Clothing R_{ecl}, m$^2\cdot$kPa/W	Moisture Permeability Index i_m
#11 Football Uniform (Warm weather)	0.199 1.280	0.123 0.795	0.032	0.022	0.38
#12 Football Uniform (Temperate Weather)	0.228 1.467	0.152 0.982	0.038	0.028	0.36
#14 Practice Football Ensemble (Hip Girdle)	0.190 1.222	0.111 0.717	0.030	0.020	0.38
#15 Practice Football Ensemble (Shorts)	0.178 1.145	0.099 0.640	0.027	0.017	0.40
#21 Basketball Uniform	0.142 0.918	0.058 0.372	0.021	0.011	0.42
#22 Basketball Uniform (with Shirt and Compression Shorts)	0.161 1.038	0.077 0.497	0.024	0.015	0.40
#31 Soccer Uniform	0.155 0.998	0.073 0.470	0.021	0.012	0.44
#41Women's Aerobic/ Running Ensemble	0.107 0.688	0.018 0.116	0.014	0.004	0.45
#42 Long Distance Running Ensemble	0.125 0.808	0.039 0.252	0.018	0.008	0.42
#43 Casual Running Ensemble	0.147 0.946	0.063 0.405	0.020	0.010	0.45
#44 Running Ensemble (Cool Weather)	0.150 0.970	0.068 0.438	0.021	0.011	0.44
#51 Men's Tennis Ensemble	0.157 1.013	0.074 0.476	0.023	0.013	0.42
#52 Women's Tennis Ensemble	0.128 0.828	0.042 0.272	0.021	0.011	0.38
#61 Cycling Ensemble	0.144 0.927	0.062 0.399	0.018	0.009	0.47
#62 Cycling Ensemble with Shirt	0.182 1.172	0.101 0.649	0.026	0.017	0.43
#71 Golf Ensemble	0.171 1.103	0.089 0.576	0.027	0.018	0.38
#81 Baseball Uniform (Warm Weather)	0.174 1.124	0.099 0.635	0.026	0.018	0.40
#82 Baseball Uniform (Temperate Weather)	0.193 1.247	0.118 0.762	0.030	0.022	0.39

The evaporative resistance value for the still air layer (R_{ea}) was 0.012 m²·kPa/W. The intrinsic evaporative resistance values for the sports ensembles ranged from 0.004 m²·kPa/W to 0.028 m²·kPa/W. The football uniforms (#11, #12) and baseball uniform worn in temperate weather had evaporative resistance values that were several times greater than that of the women's aerobic/running ensemble (#41).

Discussion

This paper presents data for the thermal insulation and evaporative resistance of selected sports apparel ensembles. These thermal parameters affect the heat exchange between the body, the clothing, and the environment. These data can be used to model the heat balance and heat storage of players during games and practices. These data were collected under still air conditions with a stationary manikin. During sports activity, higher levels of air velocity and body motion will increase the convective and evaporative heat loss from the player's body. Rational models can incorporate such situation-specific variables. Data for the ensembles tested in this study can be compared to data for other clothing ensembles in the literature [7].

Ensembles worn for a given sport often vary with respect to the amount of body surface covered by garments, the number of fabric layers on different parts of the body, the looseness or tightness of fit, the type of garments to be worn (e.g., knee-length socks versus ankle-length socks), seasonal differences, and the thickness and permeability of the component fabrics. Their thermal resistance and evaporative resistance values change with these variations. The differences are even greater when ensembles worn for different sports are compared.

A high insulation value may or may not be desirable in sports apparel, depending upon the environmental conditions in which the game is played and upon the energy expenditure of the athlete. Clothing insulation increases as the amount of body surface area covered by garments increases and as the thickness of the garment layers increases. This is why the football uniforms provided the highest insulation values in the data set. The temperate weather uniform (#12) had an intrinsic clothing insulation value of about 1 clo – the insulation associated with a heavy, three-piece men's business suit. The baseball uniform for temperate weather (#82) also covered a large amount of the body surface, but it did not have thick pads; its intrinsic clothing insulation was about 25 % less than the insulation of the football uniform for temperate weather (#12).

The higher the evaporative resistance value, the less permeable the clothing is to moisture transfer. Low evaporative resistance values are usually associated with clothing comfort in both hot and cold environments. The evaporative resistance of an ensemble depends upon the moisture permeability characteristics and wicking properties of the component materials used in the clothing and the amount of skin surface that is covered by the materials. Fiber content has little to do with the moisture permeability of textiles. The openness of the fabric structure and the type of surface treatments have more of an impact. Hard solid materials (used in helmets) and foam padding, with or without vinyl covering, particularly impede the evaporation of sweat. Therefore, the sports ensembles that covered the smallest amount of the body surface (and offered the least protection) provided the least resistance to the evaporation of sweat from the body surface (e.g., #41 and #42 running ensembles). The football uniform for temperate weather (#12) had the highest evaporative resistance value because the garments covered a large portion of the body, and the impact protection materials were thick and impermeable.

The higher the permeability index, the greater the moisture permeability of the clothing, relative to its level of insulation. Consequently, high permeability index values are desirable for comfort. The i_m values measured with a manikin generally range from about 0.5 for very permeable clothing or clothing that does not cover much of the body surface to about 0 for nonpermeable clothing. Values for most indoor clothing range from approximately 0.34–0.41 [7]. The sports ensembles also had permeability indexes within this range.

Implications

Sports uniforms that cover a large portion of the body surface or provide impact protection substantially reduce heat loss from the body. There is a need for quantitative information that exercise scientists, coaches, team physicians, and athletic trainers can use to prevent heat related injuries and deaths associated with certain sports. Mathematical models for estimating the heat exchange between the body and the environment are widely available, and user-friendly computer programs could be developed. The data presented in this paper can be used in the solution of heat balance equations to predict physiological responses of athletes when wearing uniforms under different conditions.

References

[1] Murphy, R. J., "Heat Illness in the Athlete," *American Journal of Sports Medicine*, Vol. 12, No. 4, 1984, pp. 258–261.

[2] Adams, W. C., Fox, R. H., Fry, A. J., and MacDonald, I. C., "Thermoregulation During Marathon Running in Cool, Moderate and Hot Environments," *Journal of Applied Physiology*, Vol. 38, No. 6, 1975, pp. 1030–1037.

[3] Shephard, R. J., "Minimizing the Practical Problems of World-Wide Soccer Competition: Management of Heat Exposure and a Shift in Circadian Rhythms," *Science and Sports*, Vol. 14, No. 5, pp. 248–253.

[4] Murphy, R. J., "Heat Problems in the Tennis Players," *Clinics in Sports Medicine*, Vol. 7, No. 2, 1988, pp. 429–434.

[5] Cunningham, C., "Bicycling Injuries & Safety," *Emergency*, Vol. 28, No. 5, 1996, pp. 34–37.

[6] McCullough, E. A., Jones, B. W., and Huck, J., "A Comprehensive Database for Estimating Clothing Insulation," *ASHRAE Transactions*, Vol. 91, 1985, pp. 29–47.

[7] McCullough, E. A., Jones, B. W., and Tamura, T., "A Data Base for Determining the Evaporative Resistance of Clothing," *ASHRAE Transactions*, Vol. 95, Pt. 2, 1989, pp. 316–328.

[8] ISO, "ISO 7793 Hot Environments: Analytical Determination and Interpretation of Thermal Stress Using Calculation of Required Sweat Rate," International Organization for Standardization, Geneva, 1995, pp. 1–20.

[9] ASTM Standard D1777-96 (2002): Test Method for Thickness of Textile Materials, *Annual Book of ASTM Standards*, ASTM International, West Conshohocken, PA, 2002.

[10] ASTM Standard F 1291: Test Method for Measuring Thermal Resistance of Clothing Using a Heated Manikin, *Annual Book of ASTM Standards*, ASTM International, West Conshohocken, PA, 2002.

[11] ISO, "ISO 9920, Ergonomics of the Thermal Environment: Estimating of the Thermal Insulation and Evaporative Resistance of a Clothing Ensemble," International Organization for Standardization, Geneva, 1995, pp. 1–54.

[12] Woodcock, A. H., "Moisture Transfer in Textile System, Part I," *Textile Research Journal*, Vol. 32, 1962, pp. 628–633.

SECTION 4: STANDARDS AND TEST METHOD DEVELOPMENT

Journal of ASTM International, January 2005, Vol. 2, No. 1
Paper ID JAI12100
Available online at www.astm.org

Huantian Cao,[1] *Donna H. Branson,*[1] *Jinhee Nam,*[1] *Semra Peksoz,*[1] *and Cheryl A. Farr*[1]

Development of a Cooling Capability Test Method for Liquid-Cooled Textile System

ABSTRACT: A test method to evaluate the cooling capability of a 30.5 cm × 30.5 cm or smaller liquid-cooled textile system used in liquid cooling garment development has been developed. The equipment used in this method is the sweating guarded hotplate, which is used primarily to measure thermal and evaporative resistance of clothing materials in accordance with ASTM F 1868. Using the sweating guarded hotplate, a constant heat flux from the hotplate to cold liquid can be obtained. Using the conductive heat transfer equation for data analysis, this test method provides consistent and reliable evaluation of the cooling capability over a wide range of cold liquid temperatures. This method can differentiate the cooling capabilities of alternative textile systems that might vary by fabric composition, tubing material, total tubing length, and other variables of interest. Based on the zero-intercept linear regression, the thermal resistance of the textile system can be estimated, and the cold temperature requirement for a target cooling capability can be predicted.

KEYWORDS: cooling capability, liquid cooling garment, liquid-cooled textile system, sweating guarded hotplate, heat flux

Introduction

Workers in hazardous environments have to wear protective clothing and suffer intensive thermal stress because of the insulation and moisture impermeability properties of the protective clothing and heavy work in hot conditions [1–3]. A liquid cooling garment (LCG) provides an effective solution to reduce the thermal stress and makes personnel in protective clothing more comfortable and more efficient in work [4–8]. In an LCG, cool liquid is circuited inside tubes embedded in garments with the help of a battery-powered pump. Once the liquid is warmed by the body, it is circulated to a heat sink or cooler, where it is recooled [9]. There are many factors, such as tubing characteristics, liquid flow pattern, and fit, in LCG design and development that affect the efficiency and performance of an LCG and thus affect the comfort of the LCG wearer [4]. To appraise the performance of an LCG, either a human subject test or thermal manikin test is conducted.

Human subject tests were widely used in the development and evaluation of LCGs [5,7,8]. In these tests, human subjects wearing an LCG and protective clothing performed a series of exercises in an environmental chamber. Their physiological data, such as heart rate, rectal temperature, body surface temperature, sweat loss, and weight loss, were recorded. Sometimes metabolic data including oxygen consumption, carbon dioxide production, and respiratory exchange ratio were also obtained [8]. Subjective human comfort information is also available from some human subject tests [4]. However, different people in an LCG react differently

Manuscript received 7 November 2003; accepted for publication 15 June 2004; published January 2005. Presented at ASTM Symposium on Performance of Protective Clothing: Global Needs and Emerging Markets: 8th Symposium on 13 January 2004 in Tampa, FL.

[1]Assistant Professor, Professor, Graduate Student, and Associate Professor, respectively, Oklahoma State University, Department of Design, Housing and Merchandising.

physiologically to heat stress due to metabolic rate, body composition, and acclimatization. Even for the same person, body temperature and internal heat stores may vary with time of the day [4]. Therefore, in addition to the LCG, the human subject is another factor that affects testing data and may impart a bias in the LCG evaluation. It is also difficult to measure heat flux from the body to the LCG in human subject tests [10].

The thermal manikin is an important tool in the evaluation of thermal performance of clothing and provides a valid complementary tool to human subject tests [6,10]. The thermal manikin is electrically heated and divided into several individually controlled segments. Compared with human subject tests, the thermal manikin provides a more reproducible method. It can also provide input values for thermal modeling and prediction of safe and comfortable working conditions. Nevertheless, there are limitations to thermal manikin measurement. Manikin data at best represent the performance of clothing under specified test conditions, and factors such as size, fit, and posture affect test data significantly [10]. The testing results apply to the "manikin man" – the average person. However, there is no standard definition of an average person, and thermal manikin size differs from lab to lab. An LCG may fit very well on one thermal manikin but poorly on another. Since the size and fit are so important to thermal manikin test results, the test results may not be repeatable for thermal manikins in different labs. A thermal manikin is also expensive and complicated, and there are few manikins available worldwide.

In LCG design and development, it is important to know the thermal performance of the liquid-cooled textile system with various fabric compositions, tubing material, and tubing layout design, before the whole garment is developed. However, the whole garment has to be developed for either human subject tests or the thermal manikin test. Neither test can provide useful thermal performance information on a liquid-cooled textile system planned for LCG development.

The purpose of this study is to develop a cooling capability test method that tests small size (30.5 cm x 30.5 cm or smaller) liquid-cooled textile systems, rather than a whole LCG. This permits evaluation of a textile system before garment development, fabrication, and fit of garment on specific human size or thermal manikin size occur. The equipment used in this test method is less expensive and more widely available than the thermal manikin. This test method also requires less testing time than a thermal manikin test and provides repeatable and reliable test results.

Experiment

Materials

Liquid-cooled textile systems used in this research were made by thermally bonding two layers of fabric together with tubing in between them to form a fabric-tubing-fabric sandwich. Thermal adhesive Steam-A-Seam® from Warm Company, Seattle, WA, was used as the bonding agent. Two types of tubing were used in this study. One was polyvinylchloride (PVC) tubing with 3.97 mm outer diameter and 2.38 mm inner diameter. The other was 80 % polyethylene and 20 % aluminum (PE-Al) with 3.97 mm outer diameter and 2.38 mm inner diameter. Three fabrics, 100 % cotton knit with 0.69 mm thickness, 100 % cotton knit with 0.56 mm thickness, and 80 % polyester/20 % spandex knit with 0.58 mm thickness, were used in this study.

Equipment and Procedure

A sweating guarded hotplate, manufactured by Measurement Technology NW Inc., Seattle, WA, was used in this study to measure the cooling capability of 30.5 cm x 30.5 cm liquid-cooled textile systems. This equipment was developed in accordance with ASTM Test Method for Thermal and Evaporative Resistance of Clothing Materials Using a Sweating Hot Plate (ASTM F 1868). For controlled climate, the sweating guarded hotplate is placed in a Steady State & Stability Test Chamber, Model CEO910W-4, manufactured by Lunaire Limited, Williamsport, PA. The size of the sweating guarded hotplate is a 20.3 cm square test plate with 5.1 cm guard, so there are two areas of the 30.5 cm x 30.5 cm hotplate, the center area 20.3 cm x 20.3 cm test plate and edge area guard ring. Using the software associated with the sweating guarded hotplate, the heat flux between the hotplate and the textile system at both the test plate and guard ring can be obtained. Overall heat flux of the hotplate can be calculated by:

$$OAHeatflux = \frac{TPHeatflux \times TPArea + GRHeatflux \times GRArea}{TPArea + GRArea} \quad (1)$$

Here, OA is an abbreviation of the overall, TP is an abbreviation of the test plate, and GR is an abbreviation of the guard ring. The test procedure follows these steps:

1. Put the 30.5 cm x 30.5 cm liquid-cooled textile system on the hotplate, and put a layer of 1.25 cm thick non-woven fabric on top of it to minimize heat exchange between the textile system and the environment.
2. Set the environmental temperature and relative humidity in Steady State & Stability Test Chamber at 35°C and 65 %, respectively. Set the test plate, guard ring, and atmosphere temperatures of the sweating guarded hotplate at 35°C, which is approximately human skin temperature.
3. When all of the temperatures reach set points, the heat fluxes at test plate and guard ring are 0. Since there is no temperature difference between the hotplate and the environment, there is no thermal exchange.
4. Run cold water through the tubing, and the heat flow will begin.
5. After the heat flow becomes constant, measure cold water inlet temperature, outlet temperature, and flow rate.
6. After heat flow becomes constant, calculate the average heat flux for 16 min between the textile system and the hotplate. There is no heat exchange between the hotplate and environment, so all of the heat flux is due to the temperature difference between the hotplate and textile system. The unit for heat flux is W/m^2.

Cooling Capability Test Data Analysis: Conductive Heat Transfer Equation

The thermal exchange between the hotplate and the liquid-cooled textile system is through conduction, and the heat current can be determined by the conductive heat transfer equation:

$$H = \frac{A(T_h - T_c)}{R} \quad (2)$$

Here, H is heat current, A is surface area, T_h is hot plate temperature, T_c is cold water temperature, and R is thermal resistance between the hotplate and the cold liquid inside the tubing. Let $\Delta T = T_h - T_c$, and Eq 2 can be transformed to:

$$\frac{H/A}{\Delta T}=\frac{1}{R} \tag{3}$$

Here, H/A is the heat flux at the hotplate. From Eq 3, we can see that heat flux divided by ΔT should be constant for a textile system since 1/R value for a given liquid-cooled textile system should be constant. A larger $\frac{H/A}{\Delta T}$ implies better cooling capability of the liquid-cooled textile system, and the 1/R value can reflect the cooling capability.

To determine the temperature difference between the hotplate and the cold water temperature (ΔT), inlet (T_{in}) and outlet (T_{out}) temperatures of cold water were measured. Average cold water temperature (T_c) inside the tubing was calculated by:

$$T_c=\frac{T_{in}+T_{out}}{2} \tag{4}$$

Since the hotplate temperature (T_h) was set at 35°C, the temperature difference between the hotplate and the water inside the tubing was:

$$\Delta T = T_h - T_c = 35 - T_c \tag{5}$$

Results and Discussion

Feasibility Study of Cooling Capability Test Method

Figure 1 is one example of the typical heat flux at the hotplate over a period of 2 h. Before running cold water inside the tubing at 88 min, the heat flux at the hotplate was 0, and after running cold water, the heat flux started fluctuating. At the time of 100 min, the heat flux reached constant. The reason to take the average heat flux for 16 min is to have enough data for a valid estimation and a reasonable testing time. If the data are recorded every 20 s, there are 80 data points in this 16 min period, and this should be enough for a valid estimation. The average heat flux for 16 min after reaching a constant heat flux was 274.40 W/m^2 in this test. This heat flux is due to the heat exchange between the hotplate and running cold water and can reflect the cooling capability of the liquid-cooled textile system.

To determine the feasibility of this cooling capability test, two 30.5 cm x 30.5 cm liquid-cooled textile systems were made with the same tubing layout and 805 cm total tubing length, the same PVC tubing material and outer layer fabric (the fabric facing up in the test and not in contact with the hotplate), a 100 % cotton knit (0.69 mm thickness). The only difference between these two liquid-cooled textile systems was the inner layer fabric (the fabric facing down in the test and in contact directly with the hotplate). One sample used a 100 % cotton knit (0.56 mm thickness) with thermal resistance (R_{ct}) value of 0.0198 m^2°C /W, and the other sample used was a 80 % polyester/20 % spandex knit (0.58 mm thickness) with thermal resistance (R_{ct}) value of 0.0080 m^2°C /W. The thermal resistance values were measured using the same sweating guarded hotplate in accordance with ASTM F 1868. Two types of cold water, tap water at room temperature and ice-water mixture, were run inside the tubing for the test. In the test, the water was running at different flow rates to represent coolers equipped with different pump powers. The test results for the two samples are given in Tables 1 and 2, respectively.

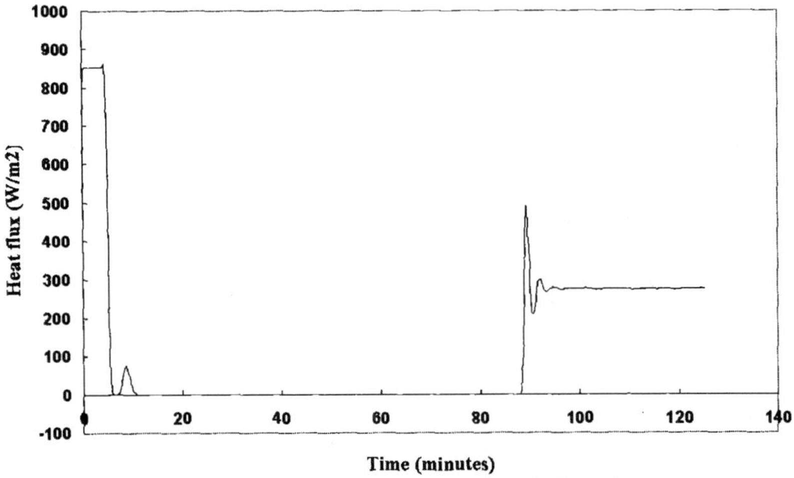

FIG. 1—*Heat flux at hotplate to test a liquid-cooled textile system.*

TABLE 1—*Cooling capability test result for liquid-cooled textile System with 100 % cotton knit as inner fabric layer.*

Flow Rate (ml/s)	ΔT (°C)	TP* H/A (W/m^2)	GR† H/A (W/m^2)	OA‡ H/A (W/m^2)	TP* 1/R (W/m^{2}°C)	GR† 1/R (W/m^{2}°C)	OA‡ 1/R (W/m^{2}°C)
2.160	12.00	242.48	165.53	199.73	20.207	13.794	16.644
2.610	12.00	257.55	179.74	214.32	21.463	14.978	17.860
3.275	12.15	255.87	177.12	212.12	21.059	14.578	17.458
4.468	11.95	262.90	183.30	218.68	22.000	15.339	18.300
6.019	12.40	273.61	189.95	227.13	22.065	15.319	18.317
8.247	13.55	278.77	194.26	231.82	20.573	14.337	17.108
10.755	12.30	275.04	190.75	228.21	22.361	15.508	18.554
12.603	13.85	310.04	216.64	258.15	22.386	15.642	18.639
5.558	27.80	631.25	437.39	523.55	22.707	15.734	18.833
5.558	27.65	630.72	442.18	525.98	22.811	15.992	19.023
...	...	...	...	Average	21.763	15.122	18.074

*TP denotes test plate, †GR denotes guard ring, and ‡OA denotes overall.

TABLE 2—*Cooling capability test result for liquid-cooled textile system with 80 % polyester/20 % spandex knit as inner fabric layer.*

Flow Rate (ml/s)	ΔT (°C)	TP* H/A (W/m^2)	GR† H/A (W/m^2)	OA‡ H/A (W/m^2)	TP[a] 1/R (W/m^{2}°C)	GR† 1/R (W/m^{2}°C)	OA‡ 1/R (W/m^{2}°C)
2.437	11.20	274.29	202.28	234.82	24.490	18.061	20.918
3.428	10.75	267.40	202.77	231.49	24.874	18.862	21.534
4.019	10.85	266.83	204.68	232.30	24.593	18.865	21.410
5.417	11.45	274.70	208.57	237.83	23.695	18.216	20.771
6.485	11.50	285.71	218.44	248.34	24.844	18.995	21.595
4.484	27.75	695.96	506.63	590.78	25.080	18.257	21.289
4.385	27.90	718.41	505.55	600.16	25.750	18.120	21.511
...	...	...	...	Average	24.799	18.482	21.290

* TP denotes test plate, †GR denotes guard ring, and ‡OA denotes overall.

From Tables 1 and 2, we can see that the heat flux is determined by the temperature difference between the hotplate and the cold water in the tubing (ΔT). A high heat flux is caused by a large ΔT. For both samples, at the test plate, the guard ring, and the whole hotplate (overall), heat flux divided by ΔT (1/R value, since $\frac{H/A}{\Delta T} = \frac{1}{R}$) is approximately constant for a wide range of ΔT, inlet water temperatures, and flow rates. Therefore, as we expected, thermal transfer in liquid-cooled textile system follows conductive heat transfer equation, and 1/R value can be used to reflect cooling capability of liquid-cooled textile system. This method is a reliable and repeatable test method. Both test plate and overall heat flux can be used in data analysis, and the user can decide to use a specific heat flux depending on the given situation. Statistical analysis t-test shows that the textile system with the 80 % polyester/20 % spandex knit as the inner layer has significantly higher cooling capabilities than the system with the 100 % cotton knit inner layer fabric at the test plate (p-value < 0.001), the guard ring (p-value < 0.001), and the whole plate (p-value < 0.001). This is due to the lower thermal resistance value of the 80 % polyester/20 % spandex knit fabric. Choosing a fabric with better thermal conductivity (lower thermal resistance) as an inner fabric layer can significantly improve the textile system's cooling capability.

Applications of Cooling Capability Test Method

Determine the Effect of Using Thermal Conductive Tubing on Cooling Capability of Liquid-Cooled Textile System—To determine whether a liquid-cooled textile system using tubing with higher thermal conductivity has better cooling capability, two textile systems with two types of tubing were compared. These two types of tubing, PVC and PE-Al, have the same outer diameter and inner diameter. The thermal conductivity of PVC tubing is 0.1214 W/m°C, and the thermal conductivity of PE-Al tubing is 0.1488 W/m°C. Both of these samples used the 80 % polyester/20 % spandex knit as the inner fabric layer. The two samples have the same tubing layout with 461 cm total tubing length in the area of 20.3 cm x 24.8 cm. Test results are in Table 3.

TABLE 3—*Cooling capability of liquid-cooled textile system using different tubing materials.*

PVC Tubing				PE-Al Tubing			
Flow Rate (ml/s)	ΔT (°C)	TP H/A (W/m^2)	TP 1/R (W/m^2°C)	Flow Rate (ml/s)	ΔT (°C)	TP H/A (W/m^2)	TP 1/R (W/m^2°C)
4.623	11.35	203.13	17.897	2.650	10.60	198.07	18.686
3.571	11.10	200.93	18.102	3.936	10.85	207.15	19.092
3.375	10.45	192.11	18.384	3.982	11.80	219.53	18.604
2.848	11.75	217.67	18.525	3.629	11.75	224.64	19.118
Average	...	...	18.227	...	...	...	18.875

In this test, since the tubing area size is 20.3 cm x 24.8 cm, which is smaller than the 30.5 cm x 30.5 cm hotplate size and the two samples have the same tubing layout, it is more suitable to analyze the test plate heat flux and 1/R value than the guard ring and overall values. Statistical analysis (t-test) of data in Table 3 shows that the liquid-cooled textile system using the tubing with higher thermal conductivity (PE-Al) has significantly higher (p = 0.008) cooling capability (1/R value) than the system using PVC tubing.

Determine the Effect of Total Tubing Length on Cooling Capability of Liquid-Cooled Textile System—To determine whether increasing the total tubing length can improve the cooling capability of a liquid-cooled textile system, two 30.5 cm x 30.5 cm samples were made as in Table 4. Both samples have 80 % polyester/20 % spandex knit as the inner fabric layer and PVC tubing. The test results are in Table 5.

TABLE 4—*Liquid-cooled textile samples with different total tubing length.*

Sample Number	Total Tubing Length (cm)
1	973
2	841

TABLE 5—*Cooling capability of liquid-cooled textile system with different total tubing length and layout design.*

Sample 1				Sample 2			
Flow Rate (ml/s)	ΔT (°C)	OA H/A (W/m^2)	OA 1/R (W/m^{2}°C)	Flow Rate (ml/s)	ΔT (°C)	OA H/A (W/m^2)	OA 1/R (W/m^{2}°C)
2.822	11.90	281.99	23.696	2.420	12.15	237.61	19.557
3.666	11.95	281.48	23.555	3.300	12.30	255.16	20.744
4.313	11.95	287.20	24.033	4.707	12.90	270.78	20.990
6.813	12.55	303.83	24.210	4.817	12.30	252.24	20.507
9.211	12.45	302.97	24.335	8.289	12.50	259.33	20.746
Average	...	...	23.966	...	...	...	20.509

The two samples in this test have different tubing layout designs. Table 4 gives the total tubing length for the whole 30.5 cm x 30.5 cm sample. Yet it was impossible to estimate precisely how much tubing was in the test plate and how much tubing was in the guard ring. Therefore, for this application with different tubing layout designs, it was better to use the overall heat flux (H/A) and 1/R values rather than the test plate data. Statistical analysis of the data in Table 5 using t-test shows that Sample 1 has significantly higher (p-value < 0.001) cooling capability value (1/R value) than Sample 2. Sample 1 has 15.7 % more total tubing length than Sample 2, and the average 1/R value of Sample 1 is 16.9 % higher than average 1/R value of Sample 2. Therefore, cooling capability (1/R value) is proportional to the total tubing length.

Determine the Effect of the Existence of an Inner Fabric Layer on the Cooling Capability of a Liquid-Cooled Textile System—In a laminated liquid-cooled textile system, there are two layers of material, an inner fabric layer and tubing between the hotplate and the cold liquid inside the tubing. To determine whether the existence of an inner fabric layer significantly lowered the cooling capability of the textile system, two samples sized 30.5 cm x 30.5 cm were made and tested. These two samples had the same tubing layout with a total tubing length of 973 cm. One sample had no inner fabric layer, and PVC tubing was thermally bonded with an outer fabric layer using Steam-A-Seam® adhesive. The other sample used 80 % polyester/20 % spandex knit as the inner fabric layer. Test results are in Table 6.

These two samples had the same tubing layout design. It is better to analyze the heat flux data at the test plate since these data are more accurate than the guard ring and overall data. Statistical analysis (t-test) of the data in Table 6 shows that the cooling capability (1/R value) of sample without an inner fabric layer was significantly higher (p-value < 0.001) than the cooling capability of the sample using 80 % polyester/20 % spandex as the inner fabric layer. These data suggest that a cooling vest with tubing that directly contacts the skin will have better cooling capability. In this test, the inner layer 80 % polyester/20 % spandex knit was not a heat-spreading fabric. A liquid-cooled textile system without an inner layer might not have better cooling capability than that with inner layer having heat-spreading capabilities.

TABLE 6—*Cooling capability of liquid-cooled textile systems with and without an inner fabric layer.*

No Inner Layer Fabric				80 % Polyester/20 % Spandex Inner Fabric Layer			
Flow Rate (ml/s)	ΔT (°C)	TP H/A (W/m²)	TP 1/R (W/m²°C)	Flow Rate (ml/s)	ΔT (°C)	TP H/A (W/m²)	TP 1/R (W/m²°C)
2.565	10.45	282.49	27.033	2.420	12.15	287.71	24.177
2.838	10.80	288.07	26.673	3.300	12.30	287.60	24.069
3.772	9.05	246.19	27.203	4.707	12.90	292.81	24.503
5.267	11.05	300.27	27.174	4.817	12.30	305.98	24.381
6.167	11.20	311.39	27.803	8.289	12.50	308.29	24.762
Average	...	...	27.177	...	...	...	24.378

Predict Cold Liquid Temperature Required for a Given Vest Cooling Goal—In LCG design and development, it is important to predict cold water temperature required for a given vest cooling goal. With a precise cold water temperature prediction, a suitable cooler can be developed for a specific application of LCG. If the predicted cold liquid temperature is higher or lower than the temperature provided by a given cooler, a redesign of the tubing layout may be necessary to increase or decrease total tubing length.

Equation 2 can be transformed to:

$$\frac{H}{A} = \frac{1}{R}(T_h - T_c) \tag{6}$$

Here, heat current H is the cooling goal for an LCG, area A is the human torso area, T_h is human skin temperature, and T_c is the cold water temperature. Based on cooling capability test data with a series of measured H/A and ΔT values, a linear regression with 0 intercept can be used to estimate the 1/R value of the liquid-cooled textile system. Transforming Eq 6, and with all other parameters known, cold water temperature T_c required for a given cooling goal of LCG can be calculated by:

$$T_c = T_h - \frac{H}{A \times \frac{1}{R}} \tag{7}$$

For example, for a cooling vest with a desired cooling goal (H) of 180 W, the human skin temperature (T_h) is approximately 35°C. The "Coppelius" sweating manikin at North Carolina State University has anatomical body dimensions of a size 40 man, with a 0.533 m^2 torso (chest, back, and lower torso) area [11]. We use 0.533 m^2 as an approximation of human torso area (A) for a size 40 man. If the liquid-cooled textile system has the same tubing layout design as the samples in Table 6, with 80 % polyester/20 % spandex knit as inner fabric layer, using the 80 % polyester/20 % spandex inner fabric layer data in Table 6, a linear regression with 0 intercept gives us the estimation of 1/R value of 24.4 W/m²°C (r^2 value for this 0-intercept linear regression is 0.888). That is, $H/A = 24.4(T_h - T_c)$. Using Eq 7 and the given values of H, A, and T_h, the required cold water temperature (T_c) for this 180 W cooling goal can be calculated as 21.2°C. That is, if the cooler can provide cold liquid with an average temperature of 21.2°C, the LCG based on this tubing layout design (973 cm total tubing length in an area of 30.5 cm x 30.5

cm) can provide 180 W cooling for a size 40 man. In actual cases where humans wear LCGs, there exists heat exchange between the cooling liquid and environment, and the heat transfer process is not homogeneous all over the torso region because of the fit of the LCG. So LCG may not have the same cooling efficiency as the liquid-cooled textile system placed on hotplate, and the cooler has to provide cold liquid with an average temperature lower than 21.2°C to achieve 180 W cooling goal in real world LCG application.

Conclusion

In this research, a testing method using the sweating guarded hotplate to measure the cooling capability of 30.5 cm x 30.5 cm or smaller liquid-cooled textile systems was developed. When cold liquid with a constant inlet temperature runs inside tubing, there is a constant heat flux from a set-temperature hotplate to the liquid-cooled textile system. Using the conductive heat transfer equation, the 1/R value (R = thermal resistance between hotplate and the cold liquid in the tubing) of liquid-cooled textile system can be obtained. This 1/R value can reflect the cooling capability of the liquid-cooled textile system. For the same liquid-cooled textile system, the 1/R value is constant over a wide range of inlet cold water temperatures, flow rates, and temperature differences between the hotplate and the cold water. Therefore, this method is a reliable and repeatable cooling capability test.

All three heat fluxes (test plate, guard ring, and overall) may be considered in data analysis. To compare different tubing layout designs, the overall heat flux can better reflect the cooling capability since it is difficult to know precisely how much tubing is exactly in the test plate portion of the instrument. To compare the same tubing layout design, the test plate heat flux can better reflect the cooling capability since heat flux measured at the test plate is more accurate.

This test method can differentiate between cooling capabilities of liquid-cooled textile systems with different inner fabric layer, tubing materials, and tubing layout designs. When a lower thermal resistance textile is used as the inner fabric layer, the liquid-cooled textile system has significantly higher cooling capability. Similarly, significantly higher cooling capability is also obtained when a higher thermal conductive tubing material is used. A liquid-cooled textile system with longer total tubing length has higher cooling capability, and the cooling capability is proportional to the total tubing length. This cooling capability test also can predict the required cold liquid temperature inside tubing for a specified cooling goal, which is important in the design of an LCG.

The sweating guarded hotplate is less expensive and more widely available than the thermal manikin. Because of its relatively small size, the environment temperature and humidity are easier to control, and thermal equilibrium between the environment and the hotplate can be achieved in less time compared with the thermal manikin test. So this test method is less time-consuming. In our experiments, one test can be finished in approximately 2 hours. A liquid-cooled textile system 30.5 cm x 30.5 cm or smaller is tested in this method. Thus, a whole garment need not be fabricated in a specific size to fit a thermal manikin.

The same LCG may exhibit large variations in cooling depending on the fit of garment, and this cooling capability cannot address this important fit issue. In some regions of LCG, the tubing density may not be homogeneous or the same as the 30.5 cm x 30.5 cm or smaller size swatch, which may result in different cooling capability. Therefore, this test method is not developed to replace thermal manikin test, but to provide an excellent pre-screening method prior to manikin test.

Acknowledgments

This research is supported under Award MIPT-106-113-2000-021 from the National Memorial Institute for the Prevention of Terrorism (MIPT) and the Office for Domestic Preparedness, U.S. Department of Homeland Security. Points of view in this document are those of the authors and do not necessarily represent the official position of the U.S. Department of Homeland Security or MIPT. The authors thank Robert Taylor of Sciperio Inc., Stillwater, OK for providing the thermal conductive (PE-Al) tubing and Steve Szczesuil of the U.S. Army Soldier & Biological Chemical Command, Natick, MA for providing the PVC tubing and for making some of the liquid-cooled textile samples.

References

[1] Bernard, T. E., "Heat Stress and Protective Clothing: An Emerging Approach from the United States," *Annals of Occupational Hygiene*, Vol. 43, No. 5, 1999, pp. 321–327.

[2] Hanson, M. A., "Development of a Draft British Standard: The Assessment of Heat Strain for Workers Wearing Personal Protective Equipment," *Annals of Occupational Hygiene*, Vol. 43, No. 5, 1999, pp. 309–319.

[3] Bishop, P., Ray, P., and Reneau, P., "A Review of the Ergonomics of Work in the US Military Chemical Protective Clothing," *International Journal of Industrial Ergonomics*, Vol. 15, 1995, pp. 271–283.

[4] Nunneley, S. H., "Water Cooled Garments: A Review," *Space Life Sciences*, Vol. 2, 1970, pp. 335–360.

[5] Nag, P. K., Pradhan, C. K., Nag, A., Ashtekar, S. P., and Desai, H., "Efficiency of a Water-Cooled Garment for Auxiliary Body Cooling in Heat," *Ergonomics*, Vol. 41, No. 2, 1998, pp. 179–187.

[6] Speckman, K. L., Allan, A. E., Sawka, M. N., Young, A. J., Muza, S. R., and Pandolf, K. B., "Perspective in Microclimate Cooling Involving Protective Clothing in Hot Environments," *International Journal of Industrial Ergonomics*, Vol. 3., No. 2, 1988, pp. 121–147.

[7] White, M. K., Glenn, S. P., Hudnall, J., Rice, C., and Clark, S., "The Effectiveness of Ice- and Freon-Based Personal Cooling System During Work in Fully Encapsulation Suits in the Heat," *American Industrial Hygiene Association Journal*, Vol. 52, 1991, pp. 127–135.

[8] Xu, X., Hexamer, M., and Werner, J., "Multi-Loop Control of Liquid Cooling Garment Systems," *Ergonomics*, Vol. 42, No. 2, 1999, pp. 282–298.

[9] Figura, S. Z., "Cooling Off," *Occupational Hazards*, Vol. 59, No. 5, May 1997, pp. 81–82.

[10] Holmer, I. and Nilsson, H., "Heated Manikin as a Tool for Evaluating Clothing," *Annals of Occupational Hygiene*, Vol. 39, No. 6, 1995, pp. 809–818.

[11] *Sweating Manikin* (n.d.), accessed June 2, 2004, from North Carolina State University, Textile Protection and Comfort Center website: http://www.tx.ncsu.edu/tpacc/comfort/sweating_manikin.html.

Journal of ASTM International, July/August 2004, Vol. 1, No. 7
Paper ID JAI12098
Available online at www.astm.org

Elizabeth A. McCullough,[1] *Jianhua Huang,*[2] *and Chil Soon Kim*[3]

An Explanation and Comparison of Sweating Hot Plate Standards

ABSTRACT: There are several standards that require the use of a hot plate apparatus for measuring the dry thermal resistance (insulation) and the evaporative resistance of fabrics: ASTM D 1518, ASTM F 1868, and ISO 11092. In addition, several National Fire Protection Association standards for protective clothing (NFPA 1951, 1971, 1977, and 1999) specify the use of this method. All of these methods measure the total resistance (i.e., the resistance of the fabric and the resistance of the surface air layer). The resistance of the air layer alone is determined by conducting a "bare plate" test. This air layer resistance is often subtracted from the total resistance to determine the fabric resistance values. Different standards use different terms for these resistances (R_{et} or R_{ef}), and they report the resistances in different units ($m^2 \cdot kPa/W$ or $m^2 \cdot Pa/W$). In addition, some methods call for guarded plates, whereas others use smaller plates, for which the experimenter must adjust the data for thick specimens to compensate for the lack of a guard. Some methods control the air velocity over the specimen, while others allow any air speed, as long as the calibration requirements on standard fabrics are met. This paper discusses the differences in test instruments, conditions, measured parameters, and units so that data from different labs can be compared and understood. In addition, data from an ASTM interlaboratory study is presented so that the repeatability and reproducibility of the different test protocols on a variety of materials can be discussed.

KEYWORDS: sweating hot plate, fabric insulation, evaporative resistance, standards

Introduction

The thermal resistance (i.e., insulation value) gives a quantitative assessment of how good a fabric system is for use in clothing worn by a sedentary person in cold weather, since radiation, conduction, and convection are the major avenues for heat loss in the cold. The higher the insulation value, the more thermal protection is provided to the wearer. However, for a person working in the cold, evaporation of sweat becomes an important means of heat loss. In addition, dry heat loss decreases with increasing ambient temperature, while evaporative heat loss (cooling) increases. Consequently, at high activity levels and/or in hot environments, the insulating value alone is insufficient for characterizing and comparing fabric and clothing systems. The resistance to evaporative heat transfer also needs to be determined for the fabric system. The types of materials used in protective clothing often impede the transfer of moisture vapor, reducing the body's ability to cool itself through the evaporation of sweat.

Manuscript received 8 January 2004; accepted for publication 10 May 2004; published July/August 2004. Presented at ASTM Symposium on Performance of Protective Clothing: Global Needs and Emerging Markets: 8th Symposium on 13 January 2004 in Tampa, FL.

[1] Professor and Co-Director, Institute for Environmental Research, Kansas State University, 64 Seaton Hall, Manhattan, KS.
[2] Post Doctoral Fellow, Institute for Environmental Research, Kansas State University, 64 Seaton Hall, Manhattan, KS.
[3] Associate Professor, Department of Textile and Clothing Design, Kyung Hee University, Yongin City, Kyunggido, 449-701 Korea.

Description of Basic Hot Plate Methods

To measure the thermal resistance or insulation value of a fabric or layered fabric system, a specimen is placed on a square plate, which is heated to a constant temperature that approximates the skin temperature of a human being (e.g., 35°C). The side of the fabric that normally faces the human body in clothing applications faces the plate. The test section in the center of the plate is surrounded by a guard heater to prevent lateral heat flow from the edges of the fabric specimen, and a bottom heater prevents the downward loss of heat from the test plate and guard section. These extra heaters force all of the heat generated in the main heater to flow through the thickness of the test specimen. The plate is housed in a chamber so the environmental conditions can be controlled carefully. The side of the specimen facing away from the heated plate is exposed to relatively calm air that is cooler than the plate. The method measures the amount of electrical power it takes to maintain the plate at a constant temperature under steady-state conditions in a cooler environment. The power is proportional to heat loss through the fabric system. Data are usually collected for at least 30 min after the system has reached steady-state.

To measure the evaporative resistance, water is fed to the surface of the plate, and a liquid barrier (e.g., untreated cellophane or microporous polytetraflouroethylene film) is placed on the plate to prevent the water from wetting the fabric. Therefore, only water vapor has contact with the fabric specimen, not liquid water. The procedures are basically the same as those for the dry tests. In isothermal tests, the air temperature is the same as the plate temperature so that there is no temperature gradient for dry heat loss. The power required to keep the plate heated in the sweating mode is related to the amount of water evaporating per unit time and the latent heat of vaporization. Although this is the best way to isolate and measure evaporative resistance, some people feel that it is not realistic, since dry and evaporative heat losses occur simultaneously from the body through the clothing layers to the environment, and there may be interactions between the dry and evaporation heat flows. In addition, it is difficult to make meaningful measurements on relatively impermeable materials because the power to the plate will become very low or zero. Therefore, one method calls for measuring the thermal and evaporative resistances under the same non-isothermal conditions. However, these results cannot be generalized to other environmental conditions.

Standards

There are several standards that require the use of a hot plate apparatus for measuring the dry thermal resistance (insulation) and evaporative resistance of fabrics. The first method, ASTM D 1518, Standard Test Method for Thermal Transmittance of Textile Materials [1], was developed by ASTM Committee D13 on Textiles. The thermal transmittance value can be used to calculate the thermal resistance. Fabric thickness can be used with the thermal transmittance value to calculate the thermal conductivity and resistivity parameters. However, these parameters are based on concepts applicable to homogeneous materials, and most fabrics are heterogeneous substances (i.e., mixtures of fiber and air; composites). Also, when there is radiant heat flow through the air spaces in a fabric, the resistance is not necessarily linear with thickness. Therefore, these parameters have limited usefulness for most textile applications, and they are not used in other hot plate standards. This standard also provides detailed tables for thermal unit conversions.

Other guarded hot plate standards such as ASTM C 177, Standard Test Method for Steady-State Heat Flux Measurements and Thermal Transmission Properties by Means of a Guarded Hot

Plate Apparatus [2], have also been developed for evaluating a wide range of materials under a wide range of conditions (e.g., building materials). Since these methods were not designed specifically for testing textiles, they will not be discussed here.

Later, ISO 11092 Textiles—Physiological Effects—Measurement of Thermal and Water Vapour Resistance Under Steady-State Conditions (Sweating Guarded Hot plate Test) was standardized [3], based on the "skin model" plate apparatus developed by scientists at the Hohenstein Institute in Germany. This standard provides methods for measuring both the intrinsic thermal resistance and the intrinsic evaporative resistance of textiles. Then the National Fire Protection Association developed a method for determining the total heat loss through a fabric assembly for use in turnout clothing, which uses a guarded hot plate under standardized conditions [4]. Later, this procedure and minimum specifications for total heat loss were added to several NFPA standards [5–8]. Meanwhile, members of ASTM F 23 Committee on Protective Clothing decided to compile the existing hot plate methods and commonly used thermal parameters into one document. This resulted in ASTM F 1868, Standard Test Method for Thermal and Evaporative Resistance of Clothing Materials Using a Sweating Hot Plate [9].

Differences and Similarities Among the Hot Plate Standards

Thermal Parameters and Nomenclature

All of these methods measure the total thermal resistance (or transmittance) of a fabric system and the surface air layer. Resistance values are reciprocal to the transmittance values and are more commonly used. The thermal resistance of the air layer alone is determined by conducting a "bare plate" test. This air layer resistance is usually subtracted from the total value to determine the intrinsic resistance value for the fabric alone. This procedure is based on the assumption that the air layer resistance measured on the bare plate is the same as the air layer resistance on the surface of a fabric. This is not always true because the heat flux from the bare plate is often greater than the heat flux from the surface of the fabric, particularly for thick fabrics, unless the temperature difference between the plate surface and the air is adjusted to compensate for the added thermal resistance of the fabric. In addition, the emissivity of the plate may not be comparable to the emissivity of some fabrics, and this may affect radiant heat flux through the air layer. However, it is easier simply to measure the air layer resistance on the bare plate than to correct it for the other factors. All of the methods except ASTM D 1518 also measure the total evaporative resistance and then determine the intrinsic fabric evaporative resistance by subtracting the resistance provided by the air layer and liquid barrier on the bare plate.

The hot plate standards use different terms for the resistance (and transmittance) values, and they report the resistances in different units. A summary of the parameters and units used in each standard is given in Table 1. The equations for determining these parameters are given in the standards [1,3,9]. A major source of confusion is that ISO 11092 defines the intrinsic thermal resistance of a fabric as R_{ct} and the intrinsic evaporative resistance as R_{et}, where "t" means textile. In ASTM F 1868 and other thermal standards, "t" means total. Therefore, the ISO R_{ct} is the same as the ASTM R_{cf}, and the ISO R_{et} is the same as the ASTM R_{ef}, where "f" refers to fabric. In addition, pascal units are used in the evaporative resistance values in the ISO standard, whereas kilopascals are used in the ASTM standard. The recent revision of ASTM F 1868 explains these differences in Part B [9].

TABLE 1—*Parameters and units used in hot plate standards.*

Parameter	Standard	Designation	Units
Thermal transmittance of fabric and air layer	ASTM D 1518	U_1	$W/m^2 \cdot K$
Thermal transmittance of air layer (bare plate test)	ASTM D 1518	U_{bp}	$W/m^2 \cdot K$
Thermal transmittance of fabric only	ASTM D 1518	U_2	$W/m^2 \cdot K$
Total thermal resistance of fabric and air layer	ASTM D 1518	none	none
	ASTM F 1868 (A,C)	R_{ct}	$K \cdot m^2/W$
	ISO 11092	none	none
Total insulation of fabric and air layer	ASTM F 1868 (D)	I_t	clo
Thermal resistance of air layer (bare plate test or apparatus constant)	ISO 11092	R_{ct0}	$m^2 \cdot K/W$
	ASTM F 1868 (A,C)	R_{cbp}	$K \cdot m^2/W$
Intrinsic thermal resistance of fabric only	ASTM D 1518	R	$m^2 \cdot K/W$
	ASTM F 1868 (A,C)	R_{cf}	$K \cdot m^2/W$
	ISO 11092	R_{ct}	$m^2 \cdot K/W$
Total evaporative resistance of fabric and air layer, including liquid barrier (isothermal)	ISO 11092	none	none
	ASTM F 1868 (B,E)	R_{et}	$kPa \cdot m^2/W$
Apparent total evaporative resistance of fabric and air layer, including liquid barrier (non-isothermal)	ASTM F 1868 (C)	R_{et}^A	$kPa \cdot m^2/W$
Evaporative resistance of air layer, including liquid barrier (bare plate test or apparatus constant)	ISO 11092	R_{et0}	$m^2 \cdot Pa/W$
	ASTM F 1868 (B,C,E)	R_{ebp}	$kPa \cdot m^2/W$
Intrinsic evaporative resistance of fabric only	ISO 11092	R_{et}	$m^2 \cdot Pa/W$
	ASTM F 1868 (B,E)	R_{ef}	$kPa \cdot m^2/W$
Apparent intrinsic evaporative resistance of fabric only	ASTM F 1868 (C)	R_{ef}^A	$kPa \cdot m^2/W$
Permeability index calculated with intrinsic resistances	ISO 11092	i_{mt}	Dimensionless
Permeability index calculated with total resistances	ASTM F 1868 (E)	i_m	Dimensionless
Total heat loss (NFPA requirement)	ASTM F 1868 (C)	Q_t	W/m^2

Test Conditions

A variety of conditions are given in the hot plate standards (Table 2). The air temperature and humidity levels have little effect on the measurement of fabric insulation under steady-state conditions for the temperature ranges normally used in these standards. These environmental parameters have a potentially greater effect on the measurement of evaporative resistance.

Isothermal sweating hot plate tests eliminate the temperature gradient for dry heat loss by maintaining the plate and air at the same temperature of 35°C. The evaporative resistance values measured under these conditions can be used in models that predict wearer comfort in different conditions. The NFPA test protocol [5–9] calls for evaluating all materials under the same conditions for both the thermal resistance and evaporative resistance tests. When the air temperature is lower than the plate temperature, this non-isothermal condition often contributes to condensation in the fabric layers and between the fabric and the plate. Therefore, the time that a specimen can stay on the plate is limited to 4 h.

TABLE 2—*Test conditions used in each standard.*

Standard	Air Temperature, °C	Relative Humidity, %	Air Velocity, m/s
Dry Plate (35°C)*			
ASTM D 1518	4.5–21.1	20–80	0.1 (calm)
ASTM F 1868A	20	65	1
ASTM F 1868C	25	65	not specified
ASTM F 1868D	20	50	1
ISO 11092	20	65	1
Sweating Plate (35°C)			
ASTM D 1868B	35	40	1
ASTM F 1868C	25	65	not specified
ASTM F 1868E	35	50	1
ISO 11092	35	40	1

*ASTM D 1518 allows for a plate temperature of 33–36°C for the thermal resistance tests.

Air velocity, airflow direction, degree of turbulence, uniformity across the plate, and method of measurement all have a major effect on total thermal and evaporative resistance values that include the resistance of the air layer on the fabric surface. Most methods control the air velocity over the specimen at 1 m/s, whereas the NFPA procedure (Part C of ASTM F 1868) allows any air speed as long as the calibration requirements on standard reference fabrics are met. Since the air layer resistances (i.e., bare plate test results) are subtracted from the total resistances prior to calculating the total heat loss for the fabric system, differences between labs due to differences in air velocity are minimized. ASTM D 1518 requires a still air condition (< 0.1 m/s) under a hood for its thermal transmittance tests. ISO 11092 requires a horizontal airflow with a 5–10 % level of turbulence. ASTM D 1868 does not specify the direction of the airflow or a level of turbulence.

The most common method for inducing turbulence is to add a shear edge descending into the horizontal airstream before it reaches the hot plate. This edge can be adjusted to increase or decrease turbulence. Leading edge effects (planarity and level) can have a huge impact on the airflow uniformity. It takes only a small discontinuity in the leading edge to alter the airflow from one side to the other. In addition, the horizontal airflow contributes to non-uniformity in air temperature or humidity (or both) across the plate, particularly when the heat flux from the plate is high (i.e., tests on the bare plate and thin fabrics). The leading edge of the hot plate is exposed to the airstream first. Therefore, the air gains heat (or humidity) as it crosses the test section. These problems are not present with vertical airflow. However, few labs have a hood capable of generating vertical flow, even though this direction best simulates real-life conditions when clothing is worn [10]. ISO 11092 and ASTM F 1868 call for measuring the air velocity 15 mm above the plate in different locations. However, measuring air velocity with hot-wire anemometers is tricky, and errors can affect the consistency of the results.

Range of Materials Relative to Plate Size

ASTM D 1518 provides a method for measuring the thermal transmittance of a fabric, batting, or layered fabric system that does not exceed 50 mm in thickness and has an intrinsic thermal transmittance between 0.7 and 14 $W/m^2 \cdot K$ (i.e., thermal resistance of 0.07 to 1.4 $m^2 \cdot K/W$). The minimum size of the test section is a 150 × 150 mm square surrounded by a guard ring that is at least 63.5 mm wide. (Note: The dimensions were changed in the 1998 revision of the standard; the original dimensions were a 254 × 254 mm test section and 127 × 127 mm guard ring. The apparatus diagram still gives these dimensions in the standard.) ISO 11092 states that the maximum thermal resistance that should be measured with the apparatus and technique described in the standard is 2 $m^2 \cdot K/W$, and the maximum evaporative resistance is 700 $m^2 \cdot Pa/W$ (0.7 $m^2 \cdot kPa/W$). The minimum size of the test section of the plate is a 200 × 200 mm square surrounded by a guard ring of 15 mm. ASTM F 1868 refers to the other two standards for information on the hot plate. The scope states that range of values measured with the hot plate should fall between 0.002 to 0.2 $K \cdot m^2/W$ for thermal resistance and 0.01 to 1.0 $kPa \cdot m^2/W$ for evaporative resistance of fabrics. The range of resistances for fabrics tested by ISO 11092 seems to be too high, whereas the range of values for fabrics tested according to ASTM F 1868 seems to be too low.

The range of fabric resistance values that can be measured accurately with a particular hot plate depends upon the size of the test section, the size of the guard section, and the environmental conditions. These factors can affect the one-dimensional heat flow through the material. For example, if a fabric is very thick, a large plate and guard section are needed to prevent heat loss from the edges of the specimen, and a low environmental temperature is needed so that the heat flux through the fabric (and corresponding power level) is sufficiently high to produce an accurate, repeatable calculation of thermal resistance. ISO 11092 uses a small guard ring, but it provides a correction factor to account for edge losses for materials thicker than 5 mm.

Some of these problems with plate-fabric relationships and measurement accuracy could be solved, in part, by requiring the testing and re-testing of specific calibration fabrics that cover the possible range of values that could be measured on typical fabrics. Interlaboratory tests could be used to determine what the thermal resistance and evaporative resistance values should be when tested under standardized conditions, as well as the reproducibility and repeatability of these tests. ISO 11092 lists a very thick reference material that can used to calibrate a hot plate with respect to dry thermal resistance. No reference material for the sweating hot plate is given. The NFPA procedures (Part C of ASTM F 1868) require measuring the thermal resistance and evaporative resistance of one, two, three, and four layers of a woven shell fabric, calculating the slope of the linear regression for each parameter, and meeting the specifications within ± 10 %. Unfortunately, four layers of the shell fabric are not as thick as some of the fabric assemblies used in fire fighter turnout gear (and evaluated by this method).

Improving Procedures for Testing Certain Types of Textiles

Hydrophilic coatings and laminates increase their vapor permeability as they become hydrated. Therefore, these materials perform better at high humidity levels. When these types of materials are placed on the liquid barrier of a sweating plate, they begin to absorb moisture from the saturated surface, and they often swell, ripple, and lift off of the plate. This phenomenon traps air between the specimen and the plate, causing a higher evaporative resistance to be measured. The ISO/TC 38/WG 17 on Physiological Properties of Textiles and Clothing is currently considering a proposal to revise ISO 11092 and improve the procedures for mounting these specimens on the plate. These procedures involve smoothing and taping the specimen to the edges of the plate during the first 15 min it is placed on the plate, which allows the sample to change shape, and adjusting the tape to flatten it out again. Alternatively, a square frame with vertical slats can be placed over the fabric to hold it down. If this procedure is used, the bare plate value also must be measured with the frame in place. Another proposed method is to hydrate coated and laminated fabrics on saturated filter paper prior to putting them on the plate.

It is also difficult to measure the thermal resistance of thin fabrics accurately because they do not adhere to the plate well, and air tends to get trapped under the fabric. Consequently, the measured resistance is erroneously high. Standard procedures for the evaluation of battings are also needed. Since battings are not used in clothing and bedding without cover fabrics on either side, they are often tested with cover fabrics appropriate for the end use. Some labs subtract off the insulation of the two fabrics and the air layer. This is necessary to calculate performance indexes such as insulation per unit thickness and insulation per unit weight. Although the insulation of the component materials is not truly additive, testing the battings alone results in higher radiant and convective heat losses (and lower insulation values).

Advantages of the Hot Plate

The hot plate simulates the heat and moisture transfer from the body surface through the clothing layers to the environment. It measures the thermal resistance (i.e., insulation value) and evaporative resistance of fabrics, battings, composite materials, and layered fabric assemblies. When the procedures are performed correctly, the method generates reproducible and repeatable results. A fairly large piece of fabric is used for a test specimen, but the same specimen can be used for the dry thermal test and the sweating test.

There are several other standard tests for measuring the water vapor diffusion through fabrics (which is basically the inverse of the evaporative resistance measured under isothermal conditions) [11]. However, none of these methods measures the dry thermal resistance, and both parameters are needed for characterizing the heat exchange between a clothed body and the environment. All of these methods provide data on *fabrics*. The results cannot be applied directly to clothing without considering the amount of body surface area covered by different fabrics, the distribution of fabric layers and air layers on the body, the looseness and tightness of fit, and the increase in surface area for heat loss. Scientists at independent laboratories have developed models for predicting the thermal resistance and evaporative resistance of clothing ensembles from hot plate data on fabrics. The Kansas State model uses hot plate data and clothing circumference data to quantify the resistance of fabric layers and air layers on different parts of the body [10]. The Hohenstein model uses hot plate data and thermal manikin data (dry test only) to calculate the evaporative resistance of a clothing ensemble based on the fraction of the body surface area that each fabric covers [12].

Fabric data can only be related to human perceptions of comfort by first estimating the thermal resistance and evaporative resistance (or vapor permeability) of an entire clothing ensemble. Alternatively, the resistances could be measured with a dry and sweating manikin, which is simply a hot plate in human form [10]. Researchers at the Hohenstein Institute have conducted several studies where they have used hot plate and manikin data to estimate the intrinsic evaporative resistance of clothing ensembles and then evaluated the comfort perceptions of human subjects wearing the clothing under different environmental conditions and levels of activity. They then developed a chart that attempts to relate evaporative resistance values of the outermost layer of clothing (e.g., rain suits) to the comfort responses of people (Table 3) [13,14]. The comfort predictions in the chart are based on the assumption that the fabric whose R_{et} value is used in the chart covers the majority of the body surface and that the rest of the garments worn are permeable. Everyday, "normal" indoor clothing generally falls in the "very good" category because the fabrics are permeable to moisture transport.

TABLE 3—*Hohenstein threshold values for intrinsic fabric evaporative resistance and clothing comfort.**

Comfort Evaluation	Intrinsic Evaporative Resistance, $m^2 \cdot$ mbar/W	Intrinsic Evaporative Resistance, $m^2 \cdot$ Pa/W
Very good	$R_{et} \leq 60 \cdot 10^{-3}$	$R_{et} \leq 6$
Good	$R_{et} = 60.1 \cdot 10^{-3}$ to $130 \cdot 10^{-3}$	$R_{et} = 6.1$ to 13.0
Satisfactory	$R_{et} = 130.1 \cdot 10^{-3}$ to $200 \cdot 10^{-3}$	$R_{et} = 13.1$ to 20
Unsatisfactory	$R_{et} > 200 \cdot 10^{-3}$	$R_{et} > 20$

* The intrinsic fabric evaporative resistance values (R_{et}) were measured according to ISO 11092. The mbar units are given in the German publications; the Pa units are used in the ISO standard [13,14].

Disadvantages of the Hot Plate

Sweating hot plates, environmental chambers, and data acquisition systems are expensive and complex to use. However, the number of hot plates used by laboratories and companies is increasing. Getting some fabrics to lie flat and stay flat during the testing is a challenge. The standard test methods are not designed to measure the impact of air layers between fabrics and between a fabric and the plate or to measure the impact of air flowing through the fabric and between air layers during the dry and sweating tests. However, moving manikin and human subject tests can measure these combined effects. In addition, the hot plate is a steady-state test, and it is difficult to use under transient conditions.

Data from the Interlaboratory Study for ASTM F 1868

An interlaboratory test of ASTM F 1868 was conducted in 2001. Samples of three fabric systems were tested in each of five laboratories under different environmental conditions (Table 4). One operator in each laboratory measured the insulation value and the evaporative resistance of fabrics according to procedures given in Parts A–E of the standard. Three replications were conducted. The NFPA calibration specified in Part C was also conducted using a new bolt of Nomex duck calibration fabric. The five laboratories were: Kansas State University, Underwriters Laboratories Inc., Navy Clothing and Textile Research Facility, W. L. Gore & Associates Inc., and North Carolina State University.

The mean thermal resistance (insulation) values and evaporative resistance values for each lab, fabric, and condition are given in Tables 5 and 6, respectively. The average thermal resistance values and average evaporative resistance values were used to calculate two performance indicators: the total heat loss value and the permeability index (Table 7). Lab 5 was not able to perform the sweating hot plate tests required in parts B and E because their lab was not capable of running the specified conditions.

TABLE 4—*Fabrics used in the interlaboratory study.*

Fabric System	Fabric Layer	Description	Weight, g/m^2
1	A*	Defender 750 yellow with Shelltite finish (Southern Mills)	253
	B	Moisture barrier: Comfort Zone 2000 (Southern Mills)	171
	C	Thermal liner: Q-9 (Southern Mills)	336
2	D	Kombat 750 natural with Super Shelltite finish (Southern Mills)	257
	E	Crosstech 2C (W. L. Gore)	169
	F	Thermal liner: Ara-lite teal (Southern Mills)	257
3	G	Marmot Membrain woven outer shell fabric with monolithic polyurethane coating (Marmot Mountain Ltd.)	183

*Layers of fabric A were used for the NFPA calibration.

TABLE 5—*Mean thermal resistance values by lab, fabric, and part of standard.*

Lab	Fabric Insulation Value R_{cf}, m^2 · °C/W					
	Fabric 1: Part C (NFPA), T_a = 25°C, RH = 65 %	Fabric 2: Part C (NFPA), T_a = 25°C, RH = 65 %	Fabric 1: Part A, T_a = 20°C, RH = 65 %	Fabric 1: Part D, T_a = 20°C, RH = 50 %	Fabric 3: Part A, T_a = 20°C, RH = 65 %	Fabric 3: Part D, T_a = 20°C, RH = 50 %
1	0.186	0.140	0.180	0.189	0.014	0.011
2	0.175	0.147	0.185	0.190	0.006	0.008
3	0.173	0.127	0.171	0.166	0.008	0.007
4	0.140	0.104	0.139	0.143	0.004	0.004
5	0.179	0.130	0.176	0.182	0.013	0.012
Lab Means	0.171	0.130	0.170	0.174	0.009	0.008

TABLE 6—*Mean evaporative resistance values by lab, fabric, and part of standard.*

Lab	Fabric Evaporative Resistance Value R_{ef}, kPa · m^2/W			
	Fabric 1: Part C (NFPA), T_a = 25°C, RH = 65 %	Fabric 2: Part C (NFPA), T_a = 25°C, RH = 65 %	Fabric 3: Part B Isothermal, T_a = 35°C, RH = 40 %	Fabric 1: Part E Isothermal, T_a = 35°C, RH = 50 %
1	0.0317	0.0167	0.0202	0.0651
2	0.0330	0.0175	0.0210	0.0654
3	0.0308	0.0163	0.0180	0.0629
4	0.0321	0.0169	0.0197	0.0509
5	0.0329	0.0164	...	...
Lab Means	0.0321	0.0168	0.0197	0.0610

TABLE 7—*Mean performance indicators by lab, fabric, and part of standard.*

Lab	Total Heat Loss Q_t, W/m^2		Permeability Index, i_m
	Fabric 1: Part C (NFPA), T_a = 25°C, RH = 65 %	Fabric 2: Part C (NFPA), T_a = 25°C, RH = 65 %	Fabric 1: Part D (Dry), T_a = 20°C, RH = 50 %; Part E (Sweating), T_a = 35°C, RH = 50 %
1	145.7	232.1	0.21
2	144.1	223.4	0.23
3	151.1	240.2	0.21
4	155.7	244.4	0.23
5	143.6	238.6	...
Lab Means	148.0	235.7	0.22

The statistical analyses were conducted using computer software designed for use with ASTM Practice E 691, "Standard Practice for Conducting an Interlaboratory Study to Determine the Precision of a Test Method." Detailed tables of these statistics are given in ASTM F 1868 [9]. According to E 691, repeatability (r) concerns the variability between independent test results obtained within a single laboratory by the same operator using the same equipment. Two test results obtained within one laboratory shall be judged not equivalent if they differ by more than the r value for that product. The r values in this study for the thermal resistance of fabrics ranged from 0.003–0.027°C · m^2/W for different fabric/condition combinations. The r values for the evaporative resistance of fabrics ranged from 0.0024–0.0075 kPa · m^2/W.

Reproducibility (R) deals with the variability between test results obtained from different laboratories. Two test results obtained by different laboratories shall be judged not equivalent if they differ by more than the R value for that material. In this study, the reproducibility limits for thermal insulation ranged from 0.010–0.061°C · m^2/W for different fabric/condition combinations. The R values for evaporative resistance ranged from 0.0024–0.0202 kPa · m^2/W.

Effect of Environmental Conditions on Hot Plate Results

ASTM F 1868 was written as a compilation of several hot plate methods. Parts A and D measure insulation at 20°C, but A requires 65 % RH, and D requires 50 % RH. Parts A and C measure insulation at 65 % RH, but A requires an air temperature of 20°C, and C specifies 25°C. Theoretically, each set of steady-state conditions should generate the same test result. The insulation values for Fabric Assembly 1 were 0.170°C · m^2/W for Part A, 0.171°C · m^2/W for C, and 0.174°C · m^2/W for D. These values are not significantly different from one another. The insulation values for Fabric 3 were 0.009°C · m^2/W for Part A and 0.008°C · m^2/W for Part D. These values are not statistically different from one another either. These results indicate that the environmental conditions given in different parts of the standard will produce essentially the same results for thermal insulation (Table 5).

The NFPA method (Part C) measures an "apparent" evaporative resistance because it is measured with a temperature gradient (for dry heat transfer) as well as a vapor pressure gradient. Therefore, evaporative resistance values for Fabric 1 measured according to Parts C and E are not equivalent.

Permeability Index Values

The permeability index (i_m) is a parameter that can be calculated from the average insulation value and average evaporative resistance (Parts D and E). These values did not differ by more than 5 % from lab to lab.

NFPA Total Heat Loss Values

Part C uses the average insulation value and average evaporative resistance value to calculate the total heat loss (Q_t) that could go through a fabric assembly under the conditions used in the test (25°C, 65 % RH). In this study, the variability in Q_t values from lab to lab was less than 5 % (i.e., maximum deviation from the mean value for all labs) for Fabric Assemblies 1 and 2. To calculate the precision statistics, individual Q_t values for each fabric replicate were needed. Therefore, the insulation value and evaporative resistance value for the Rep 1 fabric sample were used to calculate Q_{t1}; this process was repeated to calculate Q_{t2} and Q_{t3}. In this study, the repeatability limits for total heat loss values ranged from 11.9–24.5 W/m^2 for different fabrics. The reproducibility limits for total heat loss ranged from 17.3–30.6 W/m^2 for different fabrics. It is important to recognize that the variability in Q_t values will be much lower when average values for a fabric are used to calculate total heat loss.

NFPA Calibration

A new bolt of 7.5 oz/yd^2 Nomex duck was used as the calibration fabric in the interlaboratory study. Table 8 summarizes the raw data for the layers of the calibration fabric. The standard specifies the slope of the resistance values for 1, 2, 3, and 4 layers of the calibration fabric and the resistance value for 4 layers (determined from the slope). The new adjusted calibration requirements for the new bolt of Nomex duck are very similar to the requirements for the previous fabric, which are in the standard (Table 9). The new required values are slightly lower than the old.

TABLE 8—*Raw data for NFPA calibration fabric (part C).*

Lab	Bare Plate Thermal Resistance R_{cbp}, °C · m²/W	Fabric Thermal Resistance R_{cf}, °C · m²/W Layers of Nomex Duck			
		1	2	3	4
1	0.040	0.025	0.048	0.065	0.086
2	0.081	0.013	0.035	0.048	0.070
3	0.073	0.027	0.041	0.063	0.079
4	0.044	0.021	0.037	0.053	0.075
5	0.075	0.029	0.052	0.074	0.096
Lab Means	...	0.023	0.043	0.061	0.081

Lab	Bare Plate i_m*, unit less	Bare Plate AR_{ebp}, kPa ·m²/W	Fabric Apparent Evaporative Resistance Value AR_{ef}, kPa · m²/W Layers of Nomex Duck			
			1	2	3	4
1	0.72	0.0034	0.0046	0.0084	0.0135	0.0155
2	0.77	0.0064	0.0044	0.0095	0.0135	0.0192
3	0.63	0.0071	0.0035	0.0075	0.0112	0.0150
4	0.90	0.0030	0.0049	0.0092	0.0145	0.0171
5	0.75	0.0061	0.0059	0.0094	0.0147	0.0176
Lab Means	...	...	0.0047	0.0088	0.0135	0.0168

*The i_m for the liquid barrier and air layer = 0.061 (R_{ct}/R_{et}) and should be > 0.7 according to the standard.

TABLE 9—*NFPA calibration requirements (part C).*

Lab	Thermal Resistance, °C · m²/W		Apparent Evaporative Resistance, kPa · m²/W	
	Slope	R_{cf} for 4 Layers	Slope	R_{ef} for 4 Layers
1	0.0212	0.086	0.0040	0.0155
2	0.0175*	0.070*	0.0048**	0.0192**
3	0.0192	0.079	0.0038**	0.0150**
4	0.0182	0.075	0.0044	0.0171
5	0.0237*	0.096*	0.0044	0.0176
New Average	0.0200	0.081	0.0043	0.0168
New Requirements*	0.020 ± 10 % (0.018–0.022)	4 × 0.020 = 0.080 ± 10 % (0.072–0.088)	0.0043 ± 10 % (0.00387–0.00473)	4 × 0.0043 = 0.0172 ± 10 % (0.0155–0.0189)
Adjusted New Requirements	0.0195 ± 10 % (0.018–0.021)	4 × 0.0195 = 0.078 ± 10 % (0.070–0.086)	0.0043 ± 10 % (0.00387–0.00473)	4 × 0.0043 = 0.0172 ± 10 % (0.0155–0.0189)

*When the new calibration slope was determined from the thermal resistance data from all five labs, two of the labs did not meet the new requirement ± 10 %. Therefore, data from these labs were omitted, and the slope was recalculated. Four labs then met the adjusted new calibration slope (and value for four layers of duck).

**When the new calibration slope was determined from the evaporative resistance data from all five labs, two of the labs did not meet the new requirement ± 10 %. Therefore, data from these labs were omitted, and the slope was recalculated. The adjusted slope was the same since one lab was just a little high, and the other was slightly low.

Conclusions

The guarded hot plate is a valuable instrument for measuring the thermal resistance and evaporative resistance of textile materials. The resistance data can be used to calculate performance indexes such as the permeability index, total heat loss, insulation per unit thickness, and insulation per unit weight. Caution should be used in relating fabric data to clothing comfort during use without first considering clothing variables (e.g., the amount of body surface area covered by different fabrics, the distribution of fabric layers and air layers on the body, etc.), human variables (e.g., activity level), environmental conditions, and duration of use.

References

[1] ASTM Standard D 1518: Test Method for Thermal Transmittance of Textile Materials, *Annual Book of ASTM Standards*, Vol. 7, No. 1, ASTM International, West Conshohocken, PA, 2002.

[2] ASTM Standard C 177: Test Method for Steady-State Heat Flux Measurements and Thermal Transmission Properties by Means of a Guarded Hot Plate Apparatus, *Annual Book of ASTM Standards*, Vol. 4, No. 6, ASTM International, West Conshohocken, PA, 2002.

[3] ISO 11092: Textiles—Physiological Effects—Measurement of Thermal and Water Vapour Resistance Under Steady-State Conditions (Sweating Guarded Hotplate Test), International Organization for Standardization, Geneva, 1995.

[4] Gohlke, D. J., "History of the Development of the Total Heat Loss Test Method," *Performance of Protective Clothing: Sixth Volume, ASTM STP 1273,* J. O. Stull and A. D. Schwope, Eds., ASTM International, West Conshohocken, PA, 1997, pp. 190–206.

[5] NFPA 1971: Standard on Protective Ensemble for Structural Fire Fighting, National Fire Protection Association, Quincy, MA, 2000.

[6] NFPA 1977: Standard on Protective Clothing and Equipment for Wildland Fire Fighting, National Fire Protection Association, Quincy, MA, 1998.

[7] NFPA 1951: Standard on Protective Ensemble for USAR Operations, National Fire Protection Association, Quincy, MA, 2001.

[8] NFPA 1999: Standard on Protective Clothing for Emergency Medical Operations, National Fire Protection Association, Quincy, MA, 2003.

[9] ASTM Standard F 1868: Test Method for Measuring Thermal Resistance of Clothing Using a Heated Manikin, *Annual Book of ASTM Standards*, Vol. 11, No. 3, ASTM International, West Conshohocken, PA, 2002.

[10] McCullough, E. A., Jones, B. W. and Tamura, T., "A Data Base for Determining the Evaporative Resistance of Clothing," *ASHRAE Transactions*, Vol. 95, Pt. 2, 1989, pp. 316–328.

[11] McCullough, E. A., Kwon, M. and Shim, H., "A Comparison of Standard Methods for Measuring Water Vapour Permeability of Fabrics," *Measurement Science and Technology*, Vol. 14, 2003, pp. 1402–1408.

[12] Mecheels, J. and Umbach, K. H., "Thermophysiological Properties of Clothing Systems," *Melliand Textilberichte* (English edition), Part I in December, 1976, Part II in January, 1977.

[13] Umbach, K. H., "Functional Weather Protection Clothing with Good Garment-Physiology Wearing Qualities," *Melliand Textilberichte* (English edition), Vol. 67, 1986.

[14] Bartels, V. T., and Umbach, K. H., "Development and Evaluation of Protective Textile Products with Multiple Function," *Textile International Forum and Exhibition (TIFE)*, Taipei, Taiwan, 14–16 November 2001.

Journal of ASTM International, April 2005, Vol. 2, No. 4
Paper ID JAI12099
Available online at www.astm.org

Portia Yarborough[1] *and Paul Schiffelbein*[2]

Improvements in the Measurement of Cut Protection Performance: Revisions to ASTM F 1790

ABSTRACT: Workers exposed to mechanical hazards that could cause cuts or lacerations to the hands and arms have a need to understand the cut protection performance of protective apparel. Cut protection performance of clothing can be measured using CEN 388, ASTM International standard F 1790-97, and ISO 13997. The ASTM and ISO test methods employ force-distance testers to determine the cut resistance of a material when exposed to a cutting edge under specified loads. The latter two test methods provide a process for differentiating the cut protection performance of various materials such as cotton, leather, and high performance fibers.

ASTM task group F23.20.02 sought to improve the reliability and reproducibility of the F 1790-97 standard in a way that creates a global standard for measuring cut performance, demonstrates equivalence of results between different test equipment, and reduces variability in the test method. The revised test method has been balloted by the main and sub-committees for adoption.

This paper will outline the results of extensive interlaboratory studies on the test equipment, calibration method, and calibration materials. The implications of other testing aspects on the harmonization of the standards will be described. This information will be useful for end users, distributors, and manufacturers interested in measuring the cut performance of clothing.

KEYWORDS: cut resistance, cut test, CPPT, TDM, hand and arm protection, standard harmonization

Introduction

Cut protection performance of clothing can be measured using ASTM International standard F 1790-97[3] and ISO 13997[4]. These test methods are based on measuring the distance a straight blade travels across the test material under a specified load before cut-through occurs, and they provide a reliable and reproducible method for differentiating the cut protection performance of various materials. Cut resistance of a specified material is reported as the load ("rating force" or "cutting force") required on the blade that leads to cut-through in a fixed "reference distance."

The ASTM and ISO cut protection test methods came from common sources, but differ in some details. The main differences between the standard test methods are highlighted in Table 1.

Neither test method was intended to be equipment specific, although the first three entries in the table clearly derived from specific test equipment characteristics. The last two table entries are more arbitrary, but still have potential impact on test results.

Manuscript received 4 December 2003; accepted for publication 7 September 2004; published April 2005.
Presented at ASTM Symposium on Performance of Protective Clothing: Global Needs and Emerging Markets: 8th Symposium on 13 January 2004 in Tampa, FL.
[1] Research Chemist, DuPont Personal Protection, Richmond, VA, 23234 U.S.A.
[2] Statistical Consultant, DuPont Company, Wilmington, DE 19898 U.S.A.
[3] Standard Test Method for Measuring Cut Resistance of Materials Used in Protective Clothing.
[4] Protective Clothing – Mechanical Properties - Determination of Resistance to Cutting by Sharp Objects.

Harmonization of the ASTM and ISO cut protection standards has been a major concern of ASTM task group F 23.20.02 over the past several years. The task group has also worked to better understand and improve the quality of data generated by the test.

TABLE 1—*Highlights of ASTM and ISO standards for measuring cut resistance.*

ASTM F1790	ISO 13997
Normal force variable and a function of the test material coefficient of friction	Normal force constant
Blade speed sinusoidal	Blade speed constant
CPP tester	TDM-100 tester
Cut resistance calculated at 25 mm blade displacement	Cut resistance calculated at 20 mm blade resistance
Blade sharpness correction equates 25 mm and 4N	Blade sharpness correction equates 20 mm and 5N

Cut Test Equipment

NFPA cut testing methods of the 1970s form the foundation for contemporary testers. European Standard EN 388[5], which uses a circular rotating blade under fixed load, was not considered acceptable for testing materials having high cut resistance – the materials of most interest in protective apparel. The circular blade edge rapidly deteriorates in use, leading to an upward bias in measured cut protection as subsequent tests are made. The ASTM task group, confirming earlier studies [1,2], found significant dulling of straight blades after a single cut-through of both knit Kevlar® and the reference neoprene.

The ASTM standard was developed in conjunction with the CPP tester, which was substantially more sophisticated than earlier designs. Concerns with the impact of blade-sample friction on measured cut resistance led to the development of the TDM-100, which was used in developing the ISO standard. Subsequent modification of the CPP cutting arm geometry appropriately addressed the friction concerns, the final result being two similar standards and test devices. The ASTM task group has undertaken a number of large inter-laboratory studies to demonstrate the performance equivalence of the various test devices, and has included the modified CPP and the TDM-100 in the standard.

Sample Mounting

The CPP tester uses a conductive mandrel, which makes electrical contact with the blade once the sample has been penetrated. A double-sided adhesive tape is used to fix the sample to the mandrel. The TDM requires a strip of conductive tape to be mounted immediately below the sample. The cut resistance of the adhesive tape is therefore added to that of the sample on one device, but not on the other. Early studies had shown negligible cut resistance of the adhesive, so this sample mounting difference was long considered unimportant.

Concerns over the past year about the putative unimportance of the added tape led to a thorough investigation by the ASTM task group. The initial concern was that the tape increased test variability. The study included modified-arm CPP (mCPP) and TDM-100 test equipment and six laboratories. Cut tests using the tape as the sample found a 20 mm rating force of

[5] Protective gloves against mechanical risks, March 1994.

roughly 100 g. Predicted cut-through is 0.025 mm with a load of 500 g, a seemingly insignificant addition.

The additive influence of the mounting tape was studied by placing a thin copper foil strip between the sample and the mounting tape, and using electrical contact between that strip and the blade as an indicator of cut-through. The average cut-through distances (mm) of the reference neoprene (Table 2) show that the contribution of the tape is, in fact, substantial.

TABLE 2—*Influence of tape on cut-through distance of neoprene at specified loads.*

	With Tape (mm)			Without Tape (mm)		
Load	500 g	600 g	700 g	500 g	600 g	700 g
TDM-100*	33.0	24.4	17.5	19.3	13.0	9.9
MCPP*	32.0	23.1	16.8	23.6	17.3	13.2

*Type of cut tester.

The relative bias in this case is roughly 75 % for the TDM-100 and 30 % for the modified CPP. Light-weight woven Kevlar® and Spectra® showed average tape-induced bias of 200–300 %, making it quite clear that "cut-resistance" of those samples, when tested with the mounting tape, is mostly cut resistance of the tape. The tape-induced bias, however, ranges between 13 and 25 mm at 500 g, 6 and 10 mm at 600 g, and 2 and 8 mm at 700 g – hardly the hundredths of a millimeter expected!

Three "protective" glove materials were also tested with and without cutting the mounting tape. Average rating forces are shown in Table 3, which also shows the interaction between tape and sample.

TABLE 3—*Average rating forces for common glove materials showing contribution of tape.*

	Average Rating Forces (g)		Bias	
Sample type	With Tape	No Tape	(g)	(%)
Leather Glove	246	157	105	71
Cotton Glove	596	424	172	41
Kevlar® Glove	1444	1148	297	26

Relative variability of neoprene cut-through distances was also increased when the tape was included in the cut (7 versus 10 % CV), although the effect was not seen with the two woven materials (possibly due to the higher variability in cut-through distances of the woven materials).

This study dispelled the notion that the mounting tape provided an insignificant addition to cut resistance. Its substantial and variable influence led the ASTM task group to propose a modified sample mounting procedure for all test instruments. The method will specify the use of a conductive strip to be placed in between the sample material and the mounting tape.

Reference Distance and Blade Sharpness Correction

The point (4 N, 25 mm)[6] lies on the force-distance curve generated by cutting a reference neoprene sample with the CPP test instrument. The blades have a usable cutting length of approximately 50 mm, so the choice of a 25 mm reference distance and calibration point is

[6] In practice, most laboratories use masses of 400 and 500 g, instead of strictly observing 4 and 5 N forces.

logical. Load-distance data are collected that span the usable cutting length, so a 25 mm "rating force" has minimum prediction error as well.

The point (5 N, 20 mm) lies on the same force distance curve for reference neoprene cut with the CPP tester and should provide a similar anchor point for calibration. The largest obstacle to changing the ASTM method to a 20 mm reference point was the large amount of data and experience accumulated at 25 mm by fiber and glove manufacturers. There was also concern that the roughly 10 % larger forces required to cut through 20 mm would lead to more variability in reported values. An inter-laboratory study (six laboratories, eleven test materials, and duplicate determinations) was used to address that concern. Relative variability (%CV) was found not significantly different for the two reference points (median %CV approximately 11). Recommended modifications to the ASTM standard are currently undergoing balloting with a 500 g reference load and calibration points, which are cut tester specific in order to harmonize with the ISO standard.

Calibration and Calibration Materials

Calibration should minimize potential test result biases caused by variations in blade sharpness and test instruments. Calibration in both ASTM and ISO cut protection standards consists of a simple multiplicative adjustment of measured cut distances using a multiplier determined by cutting a reference neoprene sample. While the concept seems valid, the calibration in practice has been less effective than anticipated in reducing inter- and intra-laboratory variability. The ASTM task group has undertaken a number of studies recently to improve the efficacy of the calibration.

One inter-laboratory study addressed alternate calibration materials. Plain-woven fabrics of Kevlar®, Spectra®, and cotton were tested for cut resistance uniformity and their ability to reduce test variability when employed as calibration standards. Relative variability of cut-through distance of these materials was on the order of 20–30 % CV, making them unacceptable for use in calibration. Polymeric sheet materials, such as Mylar® and Cyrel®, though well controlled in thickness, required either excessive loads to cut-through and/or showed unacceptable variability. The task group will continue looking for calibration materials that are uniform, stable, and effective in use.

The task group is also investigating multi-point calibration, as well as different strategies for applying the calibration factors to the test material cut-through distances. The elimination of cutting the adhesive tape – and its non-additive impact on measured cut resistance – mentioned earlier will also improve calibration effectiveness. Uniformity of blade sharpness has been addressed through discussions with the blade manufacturer.

Conclusions

Cut protection is an important quality of gloves and other apparel items. Quantifying cut protection performance offers a tool for marketing the appropriate products to consumers. Cut performance testing also facilitates new product development and product improvement, from the level of fibers to finished goods. ASTM task group F23.20.02 has aggressively addressed a number of issues associated with cut-performance measurement, resulting in substantial improvements in the test method. Harmonization of the ASTM and ISO test methods is one product, while ongoing work will continue to reduce test variability and improve cut performance assessment. The results benefit end users, product developers, and manufacturers at all stages of the supply chain.

References

[1] Lara, J., Turcot, D., Daigle, R., and Boutin, J., "A New Test Method to Evaluate the Cut Resistance of Glove Materials," *Performance of Protective Clothing, ASTM STP 1237*, Vol. 5, J. S. Johnson and S.Z. Mansdorf, Eds., ASTM International, West Conshohocken, PA, 1996.

[2] Masse, S., Lara, J., Sirard, C., and Daigle, R., "Basic Principles Used in the Development of a New Cut-Test Machine for Standardization," *Performance of Protective Clothing, ASTM STP 1273*, Vol. 6, J. O. Stull and A. D. Schwope, Eds., ASTM International, West Conshohocken, PA, 1997.

Journal of ASTM International, September 2005, Vol. 2, No. 8
Paper ID JAI12087
Available online at www.astm.org

Serge Rebouillat[1] *and Benoit Steffenino*[1]

Cut Performance Fundamentals and Norms Harmonization

ABSTRACT: The cut performance of articles made of high performance materials is currently evaluated using several norm procedures. These tend to diverge, in their predictiveness of the "real" in use performance of the considered protective articles, when the cut resistance of the compounding fiber increases. Although relationships exist between the various cut performance levels obtained using ASTM, ISO, and EN norms, the ASTM and ISO approaches remain the most suitable methods for cut level exceeding 3, as defined by the European EN388 standard, this normalized cut resistance measurement procedure is part of the ANSI/ISEA 105-2000 guidance. Various studies, such as the effect of the cutting speed and the speed profile, the force distributions, the artefacts generated by various spot coatings, etc., aim at improving the interpretation of the measured cut level. Therewith, substantial potential improvements of the cut fundamental understanding and the norm harmonization are brought to the attention of the end-users of mechanically resistant articles made of high performance fibers, such as para-aramid.

KEYWORDS: cut mechanism, high performance fibers, knitted structures, coated textile materials, blade profiles, cut fundamentals, para-aramid, KEVLAR®

Introduction

The number of hand injuries is increasing in the domestic field and to a lesser extend in the professional area. The number of portable machines, their performance, and the "do it yourself" tendency have largely contributed to this situation. In professional activities, the awareness and the prevention regarding hand injuries have substantially contributed to contain the injury level; nonetheless the hand protection remains an area where room for improvement is substantial. Thanks to gloves made of high performance fibers, the situation is improving on a worldwide basis. General guidance about hand protection is provided in the ANSI/ISEA 105-2000.

Norms developed in 1987 in Europe and in 1997 in the USA have played an important role in classifying cut resistant materials. Glove performance can now reach the upper range of the cut scale level defined by certain norms with a substantial dexterity enhancement. Norm levels are nowadays considered more as entry criteria in the proposition of a given hand protection than as a truly predictive tool to anticipate the real cut risk where experience and expert judgment prevail. The fundamental understanding of the cut mechanism also plays an essential role to complement this approach. The test equipment, which is described in the various norms relative to cut protection, such as ASTM, EN, and ISO (American, European, and International standards, respectively), are good tools to develop a basis for the cut fundamental understanding. The equipment and related procedure limitations can be revealing in terms of the test adaptations, the equipment evolution, and the identification and the development of future generations. Two examples can be used to illustrate this argument. The development of slash cut

Manuscript received 24 November 2003; accepted for publication 4 March 2005; published September 2005.
Presented at ASTM Symposium on Performance of Protective Clothing: Global Needs and Emerging Markets: 8th Symposium on 13 January 2004 in Tampa, FL.
[1] DuPont International S.A., Geneva, P.O. Box 50, Geneva, Switzerland. E-mail: serge.rebouillat@che.dupont.com.

testing, i.e., high speed cut equipment, remains important, and the measurement of the fiber ingredient cut performance can largely improve the situation at the fiber selection stage prior to glove manufacturing.

The work presented in this paper provides and illustrates a number of situations where cut fundamentals and testing conditions are both contributing to an improved understanding of the standards limitations and potential.

There is no attempt to extrapolate the results to a real risk encountered in the field, therefore this study can not constitute a basis for risk analysis but can serve as a guidance toward an improved understanding of factors which may influence the cut performance of a protective equipment.

Cut Testing Methods and Materials

The first standardized cut test method was developed in France in 1987 (prEN 388 [1], hereafter called EN) followed by a method developed by the American Society for Testing and Materials, an American standard testing method (ASTM F 1790-97 [2], hereafter called ASTM) in 1997, and its most recent international adaptation via the International Standardisation Organization (ISO-13997 [3], hereafter called ISO). General guidance about hand protection is summarized in the ANSI/ISEA 105-2000, which relates to theses various norms.

Each testing procedure is described in the norm documents [1–3] mentioned above. The three related pieces of test equipment are depicted in Fig. 1 with a schematic representation of the blade orientation, the load force application, and the blade displacement.

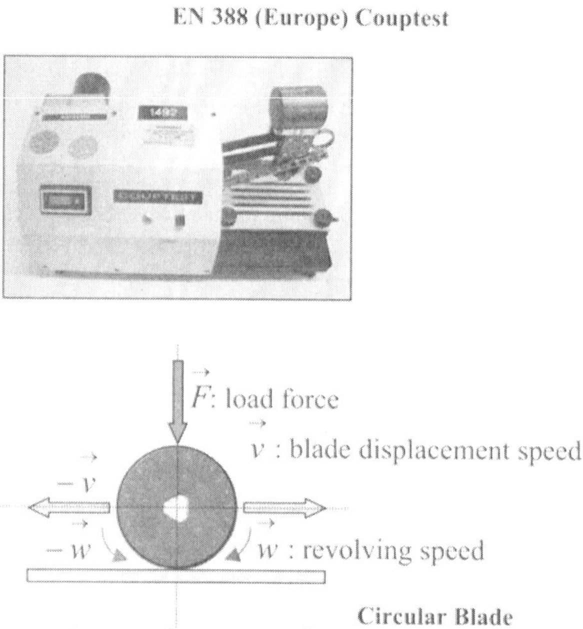

FIG. 1*a*—*Test equipments (top) with schematic representation (bottom) of the force application, (EN388).*

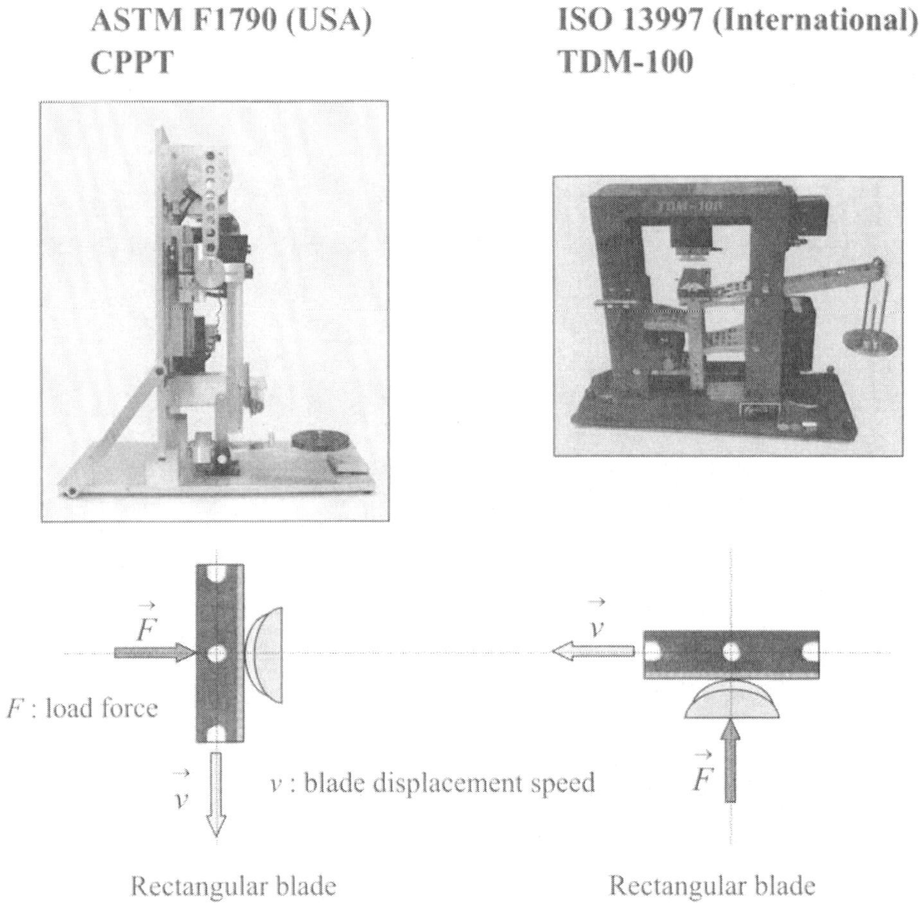

FIG. 1*b*—*Test equipments (top) with schematic representation (bottom) of the force application, blade displacement and blade orientation, (ASTM/ ISO).*

There are similarities between the ASTM and ISO procedures and equipments, but there are fundamental differences between those two and the EN equipment and test method.

The EN test method is based on the measurement of the substrate cut resistance when submitted to the cutting effect generated by the rotation of the circular blade on the sample tested, which is under a relatively moderate pressure provided by a 5 N load force applied on the blade. The number of revolutions of the blade to cause the cut-through of the specimen will be used to calculate the cut index and therewith the cut performance. A cotton fabric is used as the reference material.

The ASTM and ISO methods, on the other hand, are based on the measurement of the load necessary to provoke the cut-through of the sample after a sliding distance of the rectangular blade of 20 mm (most recently adopted harmonized distance). The corresponding load can reach, for knitted samples made of high performance fibers, the equivalent of 1000–3500 g, i.e., 10–35

N, which is 2–7 times higher than the load force applied on the circular blade in the case of the EN method. This is a difference worth highlighting.

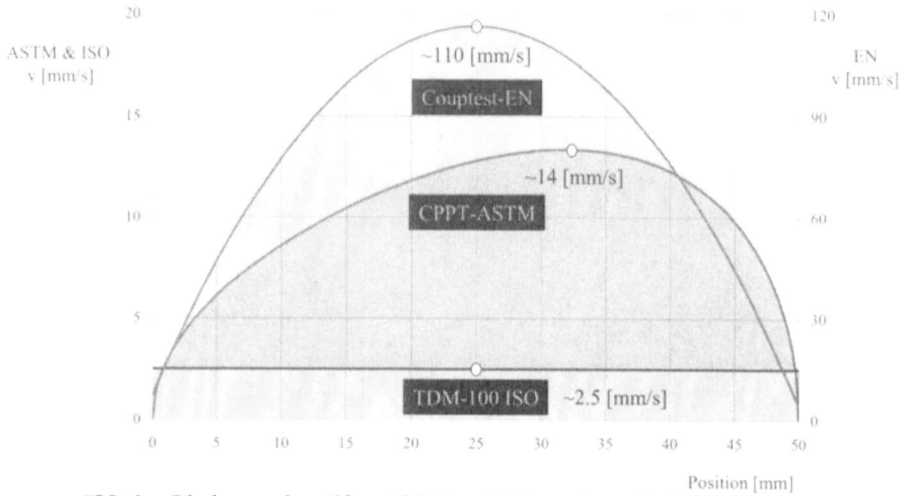

FIG. 2—*Blade speed profiles: ASTM and EN are sinusoidal, ISO is linear.*

As per Fig. 2, the blade displacements in the case of the ASTM and EN follow sinusoidal profiles with respective maximums of 14 and 110 mm/s. The ISO blade displacement profile is linear and fixed at 2.5 mm/s. A 40 fold ratio exists between the maximum 110 mm/s and the minimum 2.5 mm/s speeds. Here is another major difference, which is worth keeping in mind.

The calibration materials used for the ASTM and ISO methods differ from the one used for EN testing. Rubber versus textile is used for the former two. A key difference in the cut mechanism between the two reference materials may also be worth paying attention to, as will be illustrated later.

For all the measurements performed in this study the blades used, respectively, for the ASTM or ISO and EN tests are the following:

- ASTM and ISO: Razor Blades 88-0121, type: GRU-GRU from American Safety Razor, USA
- EN: OLFA RB 45 mm circular blades from OLFA Corporation, Osaka, Japan

The neoprene reference material used for the ASTM and ISO testing is:

- Neoprene 0.062-in. thick style NS-5550 from Fairprene, USA

The cotton fabric reference material for the EN testing is a 540 g/m^2 plain weave fabric of 1.2 mm thickness referred under:

- LEM 6 (coton écru) from ITF, Lyon, France

The double face tape used to fasten the sample specimen on the ASTM and ISO sample holder is:

- Double-faced adhesive tape, 2 in. × 36 yd, type Kendall Polyken® #100D from Can-do National Tape, USA.

Most of the materials tested in the course of this study are knitted fabrics or knitted gloves, mostly made of para-aramid fibers such as KEVLAR® staple yarns of 0.0714 g/m produced from 38 mm long segmented fibers made from continuous filament yarns composed of about 1000 filaments of 12 μm diameter. Most yarn samples were supplied by E.I. DuPont De Nemours Inc., Richmond, VA, USA.

Experimental Part

The design of the experiments was selected in order to clarify and better understand some of the aspects, which were partly introduced in the materials and methods section.

For the EN test method, the blade cutting edge degradation during the course of the testing of high cut performance materials, hereafter HCP materials, has been illustrated and studied by conducting tests on a "stop and go" basis, i.e., changing the blade during the test every 1, 2, 3, and 5 cycles as illustrated in Fig. 3. Both 100 % KEVLAR® and 5.5 % steel reinforced para-aramid knitted structures were used.

Blade replacement every (1/10 cycles)	Cn: Control specimen before (1/10 cycles)	Tn: Sample tested (1/10 cycles)	Cn+1: Control specimen after (1/10 cycles)	In : Cut index
no change	10	444	47	16.5
50	9	235	44	9.8
30	10	162	34	8.3
20	9	179	41	8.1
10	9	167	29	9.8

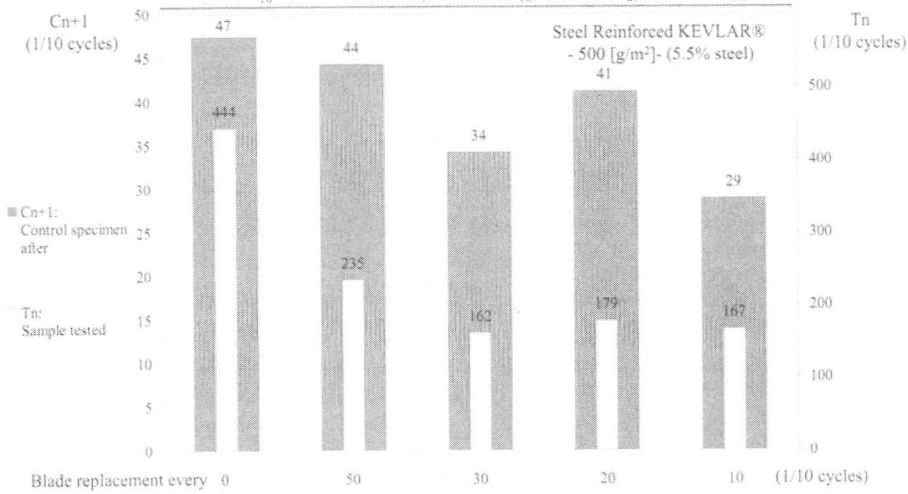

FIG. 3—*EN test method: the blade cutting edge degradation profile during the course of the testing of high cut performance materials.*

The analysis of the influence of PVC dots, melted and cured onto the knitted surface to enhance the grip properties of the fabric or the gloves, was performed using the EN and the ASTM procedures. This further illustrates some of the specific features of the two methods. Figure 4 provides the resulting data outlining NL1, which is the load, in grams, provoking the cut through of the specimen for a blade motion of 25.4 mm.

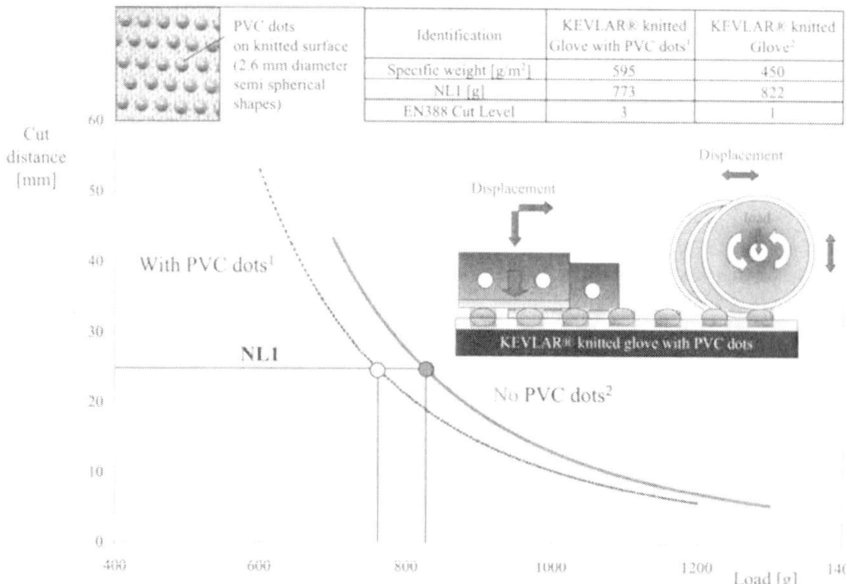

FIG. 4—*PVC dots influence on the EN (top table) and the ASTM results (top table and curves).*

The motion of the blade on the sample tested has been studied in terms of speed using the ASTM and ISO blade calibration material, i.e., a 1.6 mm neoprene layer, and 100 % para-aramid felt, whose characteristics are indicated in Fig. 5*a* and *b*.

As depicted in Fig. 6, a load cell, linked to a computer, was placed in the balanced arm holding the blade in order to estimate the force opposing the blade motion during the ASTM testing. Therewith, the ratio of this resisting force to the applied load force has been calculated for various materials, such as neoprene, 100 % KEVLAR® knitted gloves and another HCP material. Results are summarized in Fig. 7.

Figure 8 shows the modifications made to the sample holder of the ASTM equipment to simulate hot conditions. A thermal plunger was installed in the center of the sample holder. Regulation took place via a computerized interfacing module. This feasibility analysis was conducted to evaluate the expendability of the testing conditions and procedures, without attempting to duplicate actual exposure conditions.

Finally, a series of experiments was run to estimate the coefficient of variation that the user of the ASTM norm may expect when the test is repeated at constant load. A sample pre-conditioning by coating and a new sample holder, offering a tighter clamping, is proposed on Fig. 9. Using MINITAB® statistical package, the analysis of the merit of these modifications versus the traditional sample holding is proposed and illustrated in Figs. 10*a*–*d*.

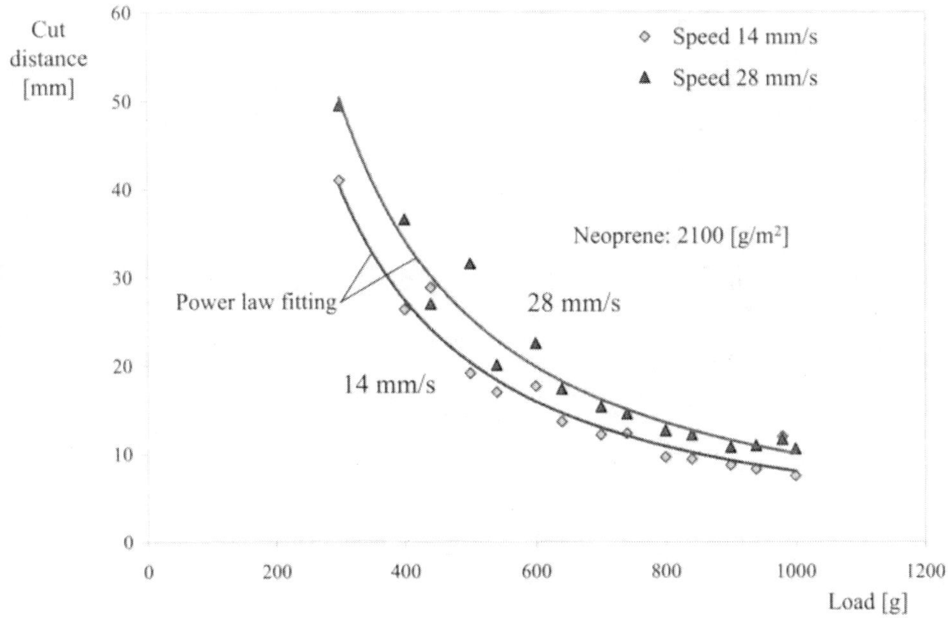

FIG. 5a—*Blade speed influence on the cut performance of the calibration material.*

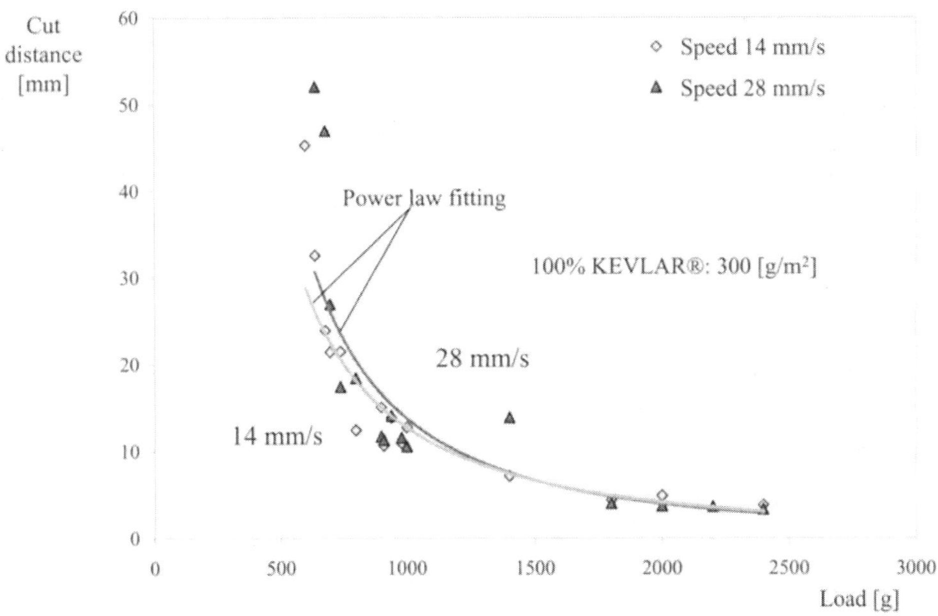

FIG. 5b—*Blade speed influence on the cut performance of a 100 % KEVLAR® textile.*

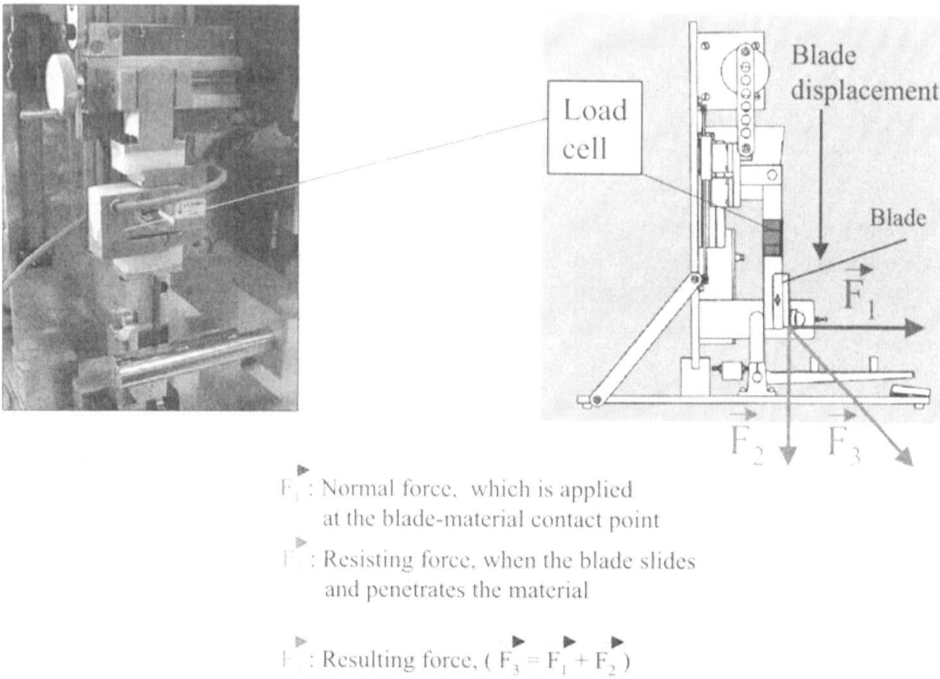

FIG. 6—*CPPT (ASTM) equipped with a compressive load cell.*

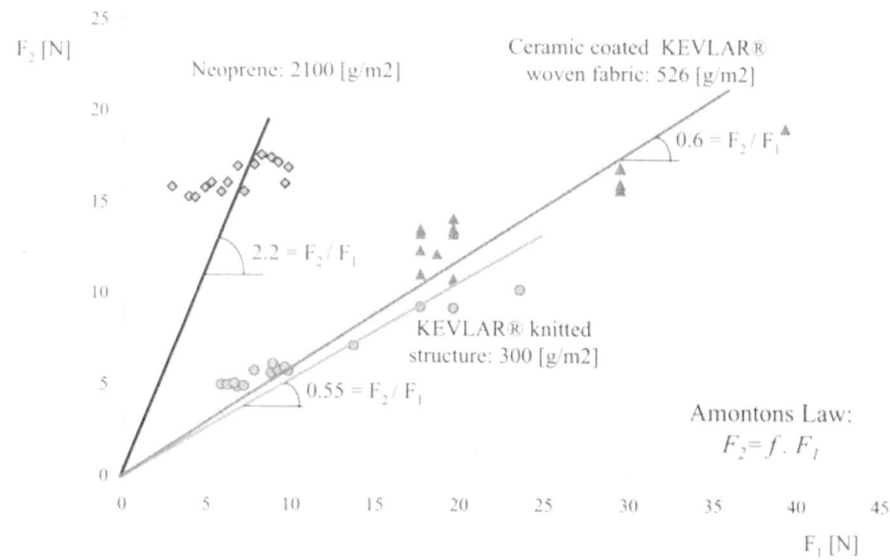

FIG. 7—*F2 / F1 ratio for various materials submitted to ASTM testing.*

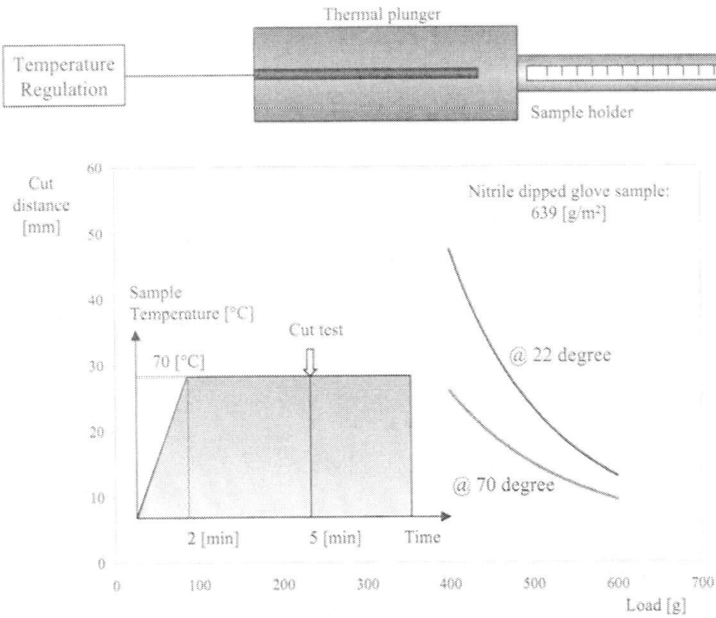

FIG. 8—*Modifications of the ASTM sample holder (top) with cut performance of a coated sample at various temperatures.*

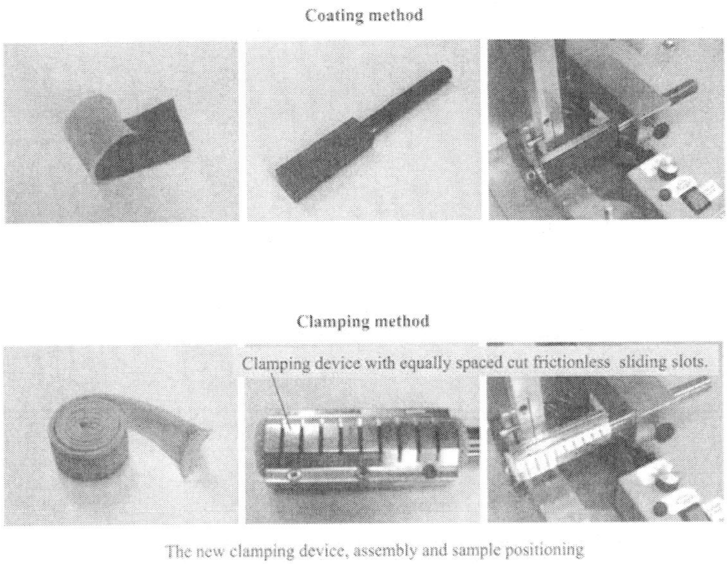

FIG. 9—*Sample pre-conditioning by coating (top) and a new sample holder adapted to the ASTM apparatus (bottom).*

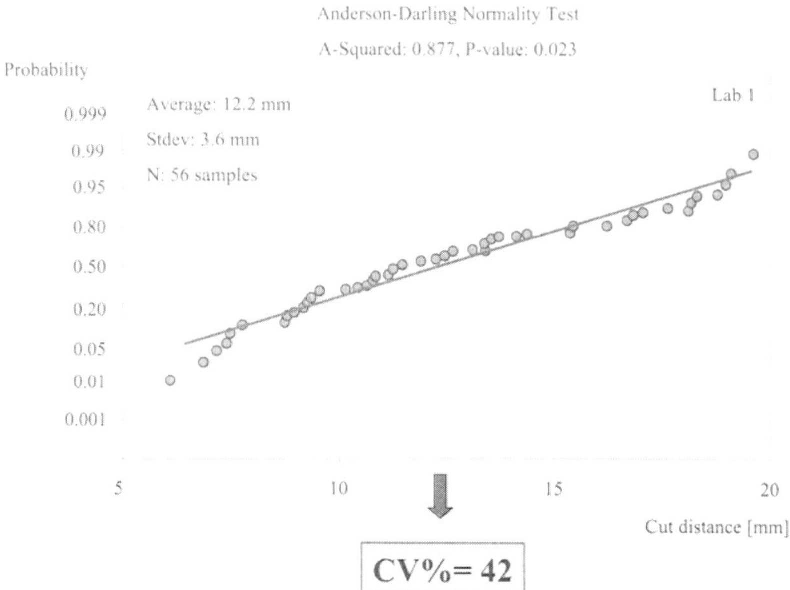

FIG. 10*a*—*Normality test on a series of samples using current ASTM; the test is repeated at constant load in Lab 1.*

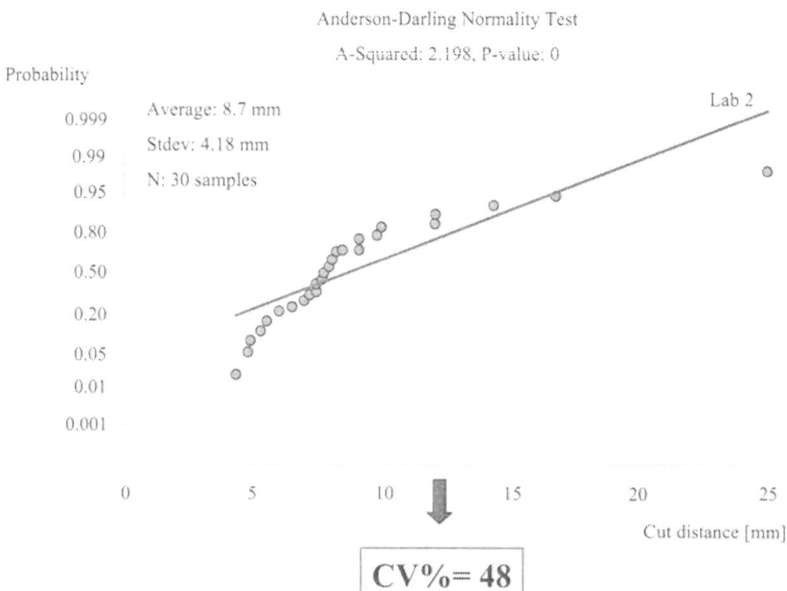

FIG. 10*b*—*Normality test on a series of samples using current ASTM; the test is repeated at constant load in Lab 2.*

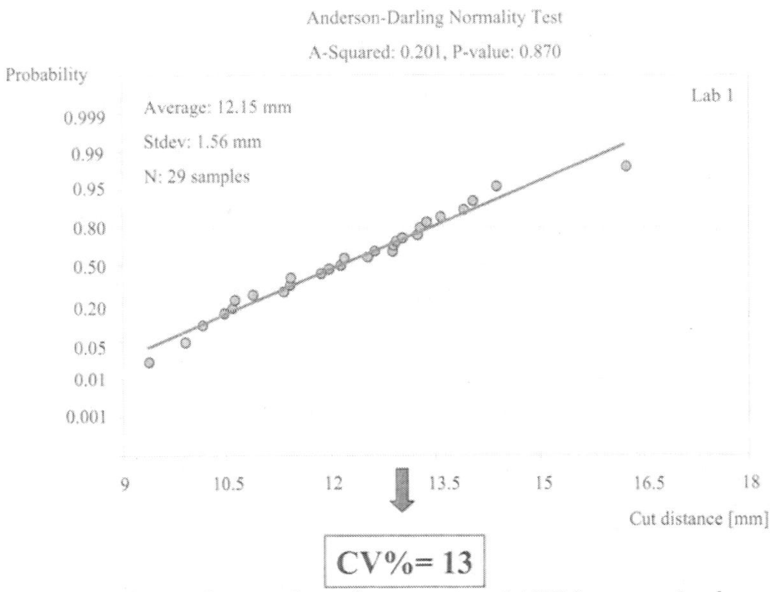

FIG. 10*c—Normality analysis performed on a series of ASTM constant load test data with sample preconditioning (coating method).*

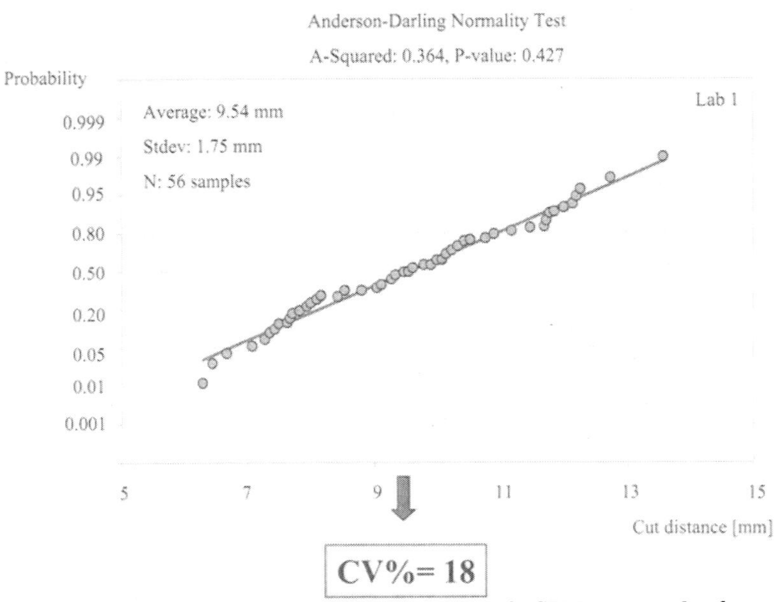

FIG. 10*d—Normality analysis performed on a series of ASTM constant load test data with new clamping device (clamping method).*

Results and Discussion

The cut mechanism and related fundamentals are rarely described in the literature. The large diversity of situations where cuts are involved and the rather large number of physical phenomena superimposed in a cutting operation render the scientific approach more difficult.

There are some qualitative reflection points which may guide the novice and help the reader through this section.

Materials which are generally considered as cut resistant materials also exhibit:

- a good tear resistance
- a reasonable puncture resistance toward sharp but not too thin perforators
- a certain resistance to abrasion or ability to fibrillate

In the cutting process itself, one may consider the following as useful considerations:

- the amount of material per unit volume, which the cutting threat has to go through
- the deformability of this material and the tendency for this material to maintain a high density under pressure, still keeping its flexibility
- the mobility of the fibers on which the blade is sliding, a revolving material being more difficult to cut.
- the blade dulling effect of the material being cut. Glass is a good example.

Fibers with skin-core structures are generally more protective. The thickness of the skin being a positive attribute. Para-aramid fibers are cut resistant. Rather brittle materials tend to be less effective. On the other hand, ductility can be a good attribute. Steel and glass tend to support these two statements. As exhibited in Fig. 11, the design of blade profiles, leaving apart the steel selection and processing, remains an art. For example, what are the criteria taken into account to make an effective knife cutting edge to cut tomatoes versus bread versus meat versus cheese, versus wood, versus textile, etc.?

When one observes a few cutting edges, it seems that, as depicted in Fig. 12, the sequence of -1- material blockage-densification and compression followed by -2- the effective sliding and cut of the previously compressed materials and -3- the relaxation and separation of the "already cut material" is a sequence which may be taken into account to understand a large proportion of cutting edge profiles. That densification-cut-relaxation sequence repeats itself along the blade as the cutting operation proceeds. The threshold compression ratio at which cut can progressively take place depends on the fiber mobility and deformation/cut speed. The latter is not taken into account in the standards procedures, except by being fixed at a given value, which differs significantly between the three ASTM, ISO, and EN standards as shown in Fig. 2. An almost 50-fold ratio exists between the maximum EN selected speed versus the ISO specified one. None of those speed values and profiles can be compared with the blade displacement profile of the knife during a criminal stabbing. Slash cut performance remains a virgin domain when hand protection equipment is specified.

Although the norms are not generally considered as truly representative of real situations, nonetheless they provide guidance. Furthermore, the analysis and identification of the norm limitations constitute a good basis for their evolution. In this study, a set of operating conditions has been analyzed. We therefore recommend following the progression of this study in the proposed sequence, although each section has been written in a way to be self-understandable.

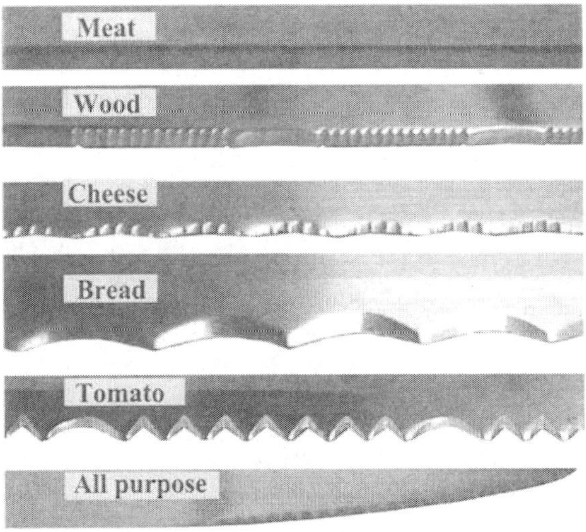

FIG. 11—*Various knife cutting edge profiles and recommended uses.*

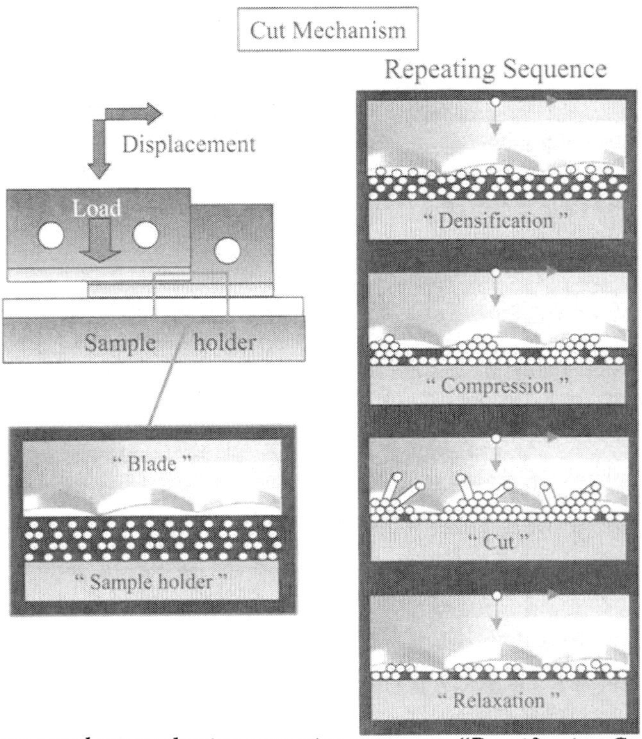

FIG. 12—*A proposed cut mechanism repeating sequence: "Densification-Compression-Cut-Relaxation-Densification..."*

Blade Degradation During EN Testing

Using the EN norm, the degradation effect of the materials on the blade cutting edge was already demonstrated by Lara [4] with a series of tests performed on fiberglass reinforced glove samples. According to this study, a fraction of a cycle – a cycle being represented by one blade full rotation counter-clockwise and its reverse rotation clockwise - was initially sufficient to cut the cotton calibration fabric. The same blade, after running for 50 revolution cycles on the HCP glass-containing material, then cut the reference cotton fabric after 25 cycles compared to 0.8 cycles initially. This shows that less than a cycle was necessary to dull the blade. Glass fibers are known for their dulling effect. A major concern shall be outlined regarding products which have a dulling effect. Shall they be considered as effective high cut performance materials?

A comparison between the cut performance of a glass containing glove sample tested via ASTM and EN procedures fully supports this question. Table 1 shows that a similar material is classified as a high cut performance material reaching level 5 on the EN scale, ranking from 1 to 5, while the same material only reaches a more moderate cut level of 1000 g as per ASTM testing. As reported on Table 1, 1100 g corresponds to a level 3 with a HCP material, such as para-aramid, which does not dull the blade.

Specimen	Glass containing glove sample	KEVLAR® glove sample
Composition	60 % Glass fiber 20% PE 20% PA	100% KEVLAR® fiber
Specific weight, g/m^2	530	670
EN cut level	5	3
C_{n+1}, cycles	10	1.1
ASTM NL1, g	1000	1100

PE : Polyethylene, PA : Polyamide

TABLE 1—*Cut testing results of two textile compositions.*

Two other HCP knitted samples were used to reproduce and further study these effects.

An HCP material tested as per EN procedure is cut through after 6 cycles if the same blade is kept all along the testing. This sample is cut after 5 cycles when the blade is changed every 1 cycle during the testing, which is equivalent to using 5 blades instead of one. Similarly, a series of experiments is run with a steel reinforced HCP glove stamped sample. On Fig. 3 one can appreciate the influence of the blade replacement on the measured cut performance of the material and the resulting calculated index given by the following relationship:

$$i_n = \frac{\overline{C}_n + T_n}{\overline{C}_n} \tag{1}$$

where i_n is the cut index given as a function of $\overline{C}_n$ and T_n. $\overline{C}_n$ is the average value of, C_n, the blade cycles measured for the cotton reference materials before, and, C_{n+1}, the number of cycles after the testing of the test material performed with the same blade. T_n is the number of cycles to cut through the test sample.

Tested via the ASTM method, the material of Fig. 3 would require a load of 3500 g for a cut to happen after a displacement of the blade of 25 mm, which is one of the highest cut performances ever recorded for a textile knitted fabric of 500 g/m^2. As shown in Fig. 3, with the EN method this material would not exceed level 4 with a single blade, while its performance would asymptotically coast at a level 3 medium performance, regardless of the number of blades used during the test. This situation is "mathematically absurd."

One shall note that the EN norm suggests using Eq 1 for a certain range of $\overline{C}_n$. In the norm, one implies that $\overline{C}_n$ should not exceed a value of 1 cycle, and in Ref. [5], Payot suggests replacing the blade when $\overline{C}_n$ reaches 2 cycles. As shown in Fig. 3, $\overline{C}_n$ is always superior to 2 regardless of the number of blades used, knowing that a blade practically cannot be changed below 1 cycle, i.e., a back and forth complete revolution.

Naturally, if all the calculations of the cut index of Eq 1 had been made keeping for C_{n+1} the initial value C_n, the cut performance of the material would be closer to the level of performance given by the ASTM approach.

One key and certainly essential difference between ASTM, ISO, and EN approaches is shown in Fig. 1*a* and *b*. A given portion of the blade is always used once and only once during the cutting process experienced by the material during the ASTM and ISO tests. This is not the case with the EN procedure where the minimum number of cycle, which is taken into account, is 1 for a level 2, meaning using at least twice a given section of the blade.

An additional curve is presented in Fig. 13, which gives the relationship between C_{n+1} and T_n for a series of HCP materials. A logarithmic fitting of the experimental data gives access to a mathematical relationship, which can then be used in Eq 1, for an "academic purpose" only. The index i_n then becomes a unique function of T_n. One can now compare on Fig. 14 the variation of i_n as per Eq 1 versus the variation of i_n as a unique function of T_n. One can conclude that the blade degradation leads to an underestimation of the cut performance of HCP materials, which tends toward an artificial cut performance level 3.

Then comes the question: shall the EN norm be only used for materials of cut performance above 3 when $\overline{C}_n$ is superior to 1, or 2? A partial answer is established by the fact that the ASTM/ISO norms are more and more often used for this type of materials. Furthermore, the EN norm recent revisions suggest the use of the ISO testing in such cases. This is a significant step and an improvement.

Speed Effect on Cut Resistance

A key conclusion of the above discussion is that the blade degradation, which occurs regardless of the selected norm procedure, affects the EN cut performance calculation but does not influence the ASTM or ISO ones.

As mentioned before, different blade speeds and blade or sample displacement profiles have been selected and recommended during the norm development phase and were adopted by the norm committees. There is not very much information shared in the literature, which provides some reasons for selecting one of the other speed and displacement characteristics.

Payot [5] mentioned that "after numerous trials, a 10 cm/s speed was chosen," referring as well to the fact that "too high a speed with too high a load will result in heating the specimen."

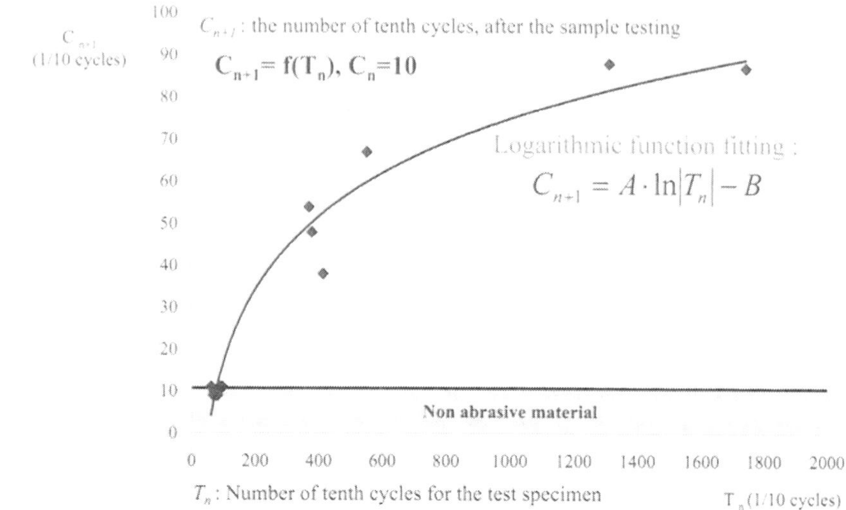

FIG. 13—*Relationship between C n+1 and Tn for a series of HCP materials.*

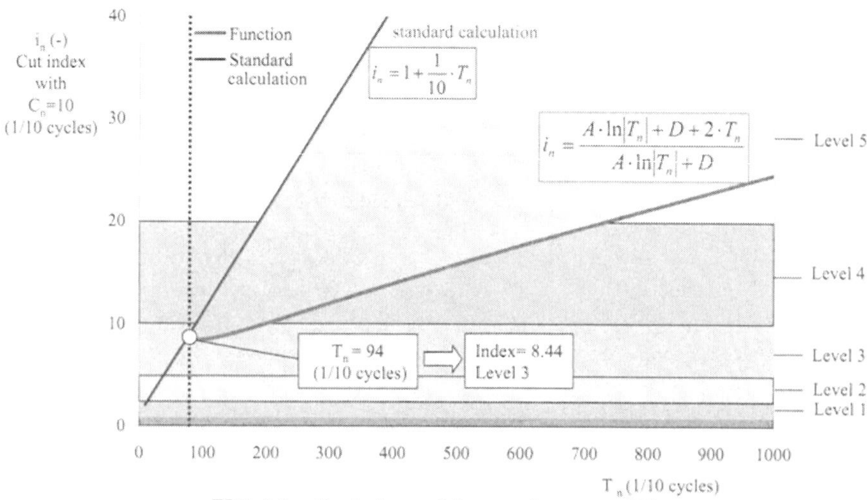

FIG. 14—*Variations of in as a function of Tn.*

Lara [7] concluded from a series of trials run at 350 mm/min and 150 mm/min, respectively, that "the effect of the blade speed on the blade displacement required to cut the materials is non-significant" and selected a blade speed of 150 mm/min "because it is easier to control."

A factor which has not been discussed is whether the blade speed can affect one material more than another. For example Fig. 5*a* and *b* compares the speed effect on a neoprene rubber layer and a KEVLAR® sample stamped out of a knitted glove for a speed variation ranging from

14–28 mm/s. Such a variation affects the neoprene load-to-blade displacement curve much more than the load-to-blade displacement relationship of the para-aramid samples. This may raise some concern about the 2.5 mm/s versus the 110 mm/s blade displacement speed respectively selected for the ISO and the EN procedures. The speed selected for the ASTM approach, i.e., 14 mm/s, and the highest speed of 28 mm/s selected in the present study lies in between the ISO and the EN selected values. Should those values be reconsidered and harmonized? This question goes beyond the scope of this study. Nonetheless, some additional data will be discussed later on and may converge with this observation providing some elements of explanation. One can already underline that in a textile structure the fibers are relatively free to move, especially when the textile is knitted, which adds bulk to the structure. This mobility can compensate for the blade displacement speed variation since the fiber elements provoke an interface slippage.

Since these standards have been put together mostly for the testing of hand protective equipment mostly made of HCP fibers such as para-aramid, the results shown on Fig. 5*b* tend to support the rather limited effect of speed within the range tested. The emergence of dipped supported gloves, however, may provoke some more questioning of the speed effect since a dipped or coated glove is no more than a textile structure covered by a rubber layer.

Coating/Printing Effect on the Cut Resistance of Textile Materials

Screen-printing is used to deposit PVC dots on textile fabrics to enhance their grip physical properties. Commercial PVCs are generally used. The screen printing takes place on carrousel stations. The resin spots are $6/cm^2$, which are generally 2.6 mm diameter semi spherical shapes.

PVC resins are considered as low modulus and low hardness materials quite deformable to ensure, in the application mentioned above, sufficient contact with a flat and rather slippery surface. PVC cannot be considered as cut resistant materials.

The testing of gloves with PVC dots is shown in Fig. 4. For the dotted side, the level 3 given by the EN cut testing is not consistent with the 770 g load necessary to provoke the cut through after a 25 mm blade displacement using ASTM method. One would expect at least a load of 1100 g, i.e., 1.5 folds, to reach an EN level 3. Furthermore, the non-coated side of the glove, i.e., without PVC dots, reaches an inferior level 1versus level 3 for the dotted side. This does not seem to make sense in terms of cut performance. For the ASTM analysis, one observes that, for the non-dotted side, the necessary load to cut through after a 25 mm blade displacement is 820 g. This is more in line with the expected results since the coating is not a cut resistant material and is a spot by spot treatment. Overall in this specific case, the EN testing largely overestimates the real cut performance.

The EN testing does not appear to reflect the real cut performance of the material. Since it tends to overestimate it, precautions shall be taken regarding the interpretation of the results. In terms of ASTM or ISO testing, if the polymeric layer deposited on the textile substrate is continuous and interlocked into this substrate, the cut performance is likely to be less than the substrate alone since the fiber's mobility is restricted. This is not the observation made from Fig. 4 since the PVC dotting is discontinuous.

No references in the literature are available on this matter except that PAYOT [5] classified rubber supported gloves as level 2 and knitted para-aramid gloves as level 3 or 4, which tends to support our conclusion that PVC dotted gloves should not exceed the performance of uncoated gloves made of the same fiber substrate.

The various tests, ASTM, ISO, and EN, reveal certain aspects of the cut mechanism, which is rather complex and multiple. Cutting with compression with or without blade motion differs significantly with cutting accompanied by slicing and sliding with blade motion. The two extremes, i.e., stab and slow abrasive cutting, are part of real situations, which make the norm specification extremely complicate.

The blade dulling, the speed effect, and the interference of non-cut performing polymeric coating have been discussed and may find some additional explanations in the study that follows.

Forces Involved During the Cutting Process

The apparatus designed to measure the cut performance as per the ASTM procedure is shown in Fig. 1 along with a schematic diagram, which provides the blade orientation and displacement and force application point. This force is generated by weights placed on a plate mounted on a lever arm assembly in a way that doubles the resulting force applied to the specimen. The test apparatus also consists of a motor-driven balanced arm, which holds the blade.

In order to measure the co-linear force component resisting to the displacement of the blade, a 50 N (or 200N) load cell, type S2 from HBM Germany, was installed in the arm holding the blade as per the diagram of Fig. 6. Connected to a computer interface, module MP55 from HBM Germany, the resisting force, $-\|\vec{F}_2\|$, was recorded during the cutting process using the CATMAN® software from HBM Germany. The normal force, $\|\vec{F}_1\|$, applied onto the blade edge is assumed to be directly proportional to the double weight force value, which is an approximation since during the displacement and due to the thickness of the sample, the blade cutting edge does not remain strictly perpendicular to the sample contacting point. This is depicted in Fig. 6. Figure 7 provides various $\|\vec{F}_2\|/\|\vec{F}_1\|$ ratios.

For the neoprene calibration material the resisting force, $-\|\vec{F}_2\|$, is about 2.2 times superior to the normal applied force, $\|\vec{F}_1\|$. The resisting force found in this study is almost equal to the one, identified as a frictional force, reported by Massé [6] in his study of the basic principles used in the development of a new cut-test machine now specified in the ISO test method. Therefore, our findings, although conducted independently on a different testing apparatus, ASTM versus ISO, are fully in line with the work conducted by Massé [6] at the IRSST, Québec, Canada. This is a good demonstration of the interchangeableness of the two pieces of equipment now both recommended in the ASTM standard.

Figure 7 also provides the $\|\vec{F}_2\|/\|\vec{F}_1\|$ ratio for a KEVLAR® knitted structure and a ceramic coated para-aramid woven fabric. The two ratios are more or less ¼ of the neoprene measured ratio, which may bring a question regarding the suitability of using such a material performance in the ASTM and ISO norms as a blade sharpness normalizing factor.

Furthermore, the amount of friction generated to cut through neoprene may explain why a dotted knitted structure exhibits an artificially high cut EN performance. The necessity to use the blade cutting edge only once during a test is essential in a situation where abrasion and resulting friction of the surface of the tested material are high.

Temperature Effect on the Cut Resistance

The sample holding mandrel was drilled to incorporate in its core a heat plunger, whose temperature is adjusted and controlled between 20–90°C via a type 2216L regulator from EUROTHERM, France.

Figure 8 shows the equipment modifications and depicts the influence of the temperature on the load to displacement relationship for a dipped knitted glove sample coated with 313 g/m^2 of commercial nitrile resin from Stockhausen, Germany, with typical FTIR bands at 2237 and 1493 cm^{-1}. The substrate is a knitted glove of 326 g/m^2 containing 48 % of polyamide fiber, 48 % of polyethylene high molecular weight fiber, and 4 % of a polyurethane elastic fiber.

The temperature, as could be anticipated from the rheological properties of the coating and the low melting temperature of the fibers, has a significant effect on its cut performance. Although quite predictable, one could consider incorporating such equipment modification and procedures in the ASTM and ISO norms to further improve the representativeness versus real situations.

Variability of the Cut Resistance of HCP Materials

To provoke a cut through of the specimen, for a given weight the blade displacement leads to a distance to load relationship, which, as illustrated in Figs. 5*a* and *b*, is non linear. The cut distance grows exponentially as the load decreases. This type of variation, as shown on these two figures, corresponds to a power law fitting therewith very different slopes, dy/dx ratios, along the curve. At low load, the precision, which one can expect on the blade displacement measurement, is much lower compared to the higher load region. Unfortunately, the adopted blade displacement, initially of 25 mm and now of 20 mm are not the most favorable to a lower variability cut testing.

Lara [1] reported coefficients of variation, CVs, ranging from 27–84 %, with an average of 42 %, for a cut distance ranking from 5.5–22 mm in the case of the ISO testing of various gloves of medium cut resistance. Tejani et al. [8] concluded that "despite variability in the cut resistance values, all laboratories ranked cut resistance of ten materials consistently." This statement tends to recognize the variability of the ASTM testing, which, as illustrated by TEJANI in the same reference, can reach 12 mm for a cut distance of 25.4 mm, i.e., 50 % variation for a load of 2500 g.

These values are not underestimated. As shown on Figs. 10*a* and *b*, the coefficient of variation, obtained at two different ASTM certified labs, reaches 42 and 48 % for a cut distance of 10 mm and a load of approximately 700 g. Furthermore, the normal probability plots are non linear, and the Anderson-Darling normality test provides respective P-values of 0.023 for one lab and 0.000 for the other, which reveals that the data are not normally distributed, although 60 data points for one lab and 30 data points for the other were gathered. The Anderson-Darling test is an alternative to the chi-square and Kolmogorov-Smirnov goodness-of-fit tests. All statistical analysis conducted in this study were done using MINITAB® statistical package.

There are numerous causes which contribute to the variability. The discussion of these goes beyond the scope of this study. We have selected one of them for the sake of conciseness and "à-propos" in terms of suggesting possible improvements of the tests discussed here.

The sample preparation and the boundary effect related to the sample holding on the mandrel of the ASTM apparatus is one of these parameters and seems to be the most effective one to reduce the CV %. Textile structures are highly deformable and composed of multiple yarn ends, which are maintained together by the knitting loop entanglement or the weaving overlapping cross-sections. The handling of such samples is delicate and can cause fabric density variation by stretching or compression under relatively low stresses.

A first attempt to reduce variability was made by coating the glove or a larger sample than the one to be tested.

A 1.2 mm thick layer, representing 360 g/m^2 of commercial nitrile coating from Stockhausen, Germany, with typical FTIR (Fourier Transformed Infrared) a bands at 2237 and 1493 cm^{-1}, is applied to the knitted sample surface and dried at 160°C for several minutes. The non-coated knitted sample weight was 320 g/m^2. The cut test is conducted as per the standard ASTM procedure. Figure 9 shows some details of the coating and testing procedure used.

This method leads to a much lower cut variability, CV % = 13, for a blade displacement of 10 mm under a load of 700 g. The normal probability plot is linear, and the Anderson-Darling normality test provides a P-value of 0.870, which supports normality. Twenty-nine data points were gathered. This was confirmed with another certified lab with a CV % of 11 for 28 data points. Figure 10*c* details these results.

A second attempt to reduce variability was made by clamping the sample and intercepting in between the textile fabric and the clamp a thin layer of neoprene. The sample preparation was designed to minimize potential distortion of the sample.

Using a circular or a regular Shima Seiki (Japan) glove knitting machine, a long tube is knitted at a diameter which is about the mandrel diameter of the cut tester. Then a given length of the knitted tube is slid over the mandrel. This is done to minimize the sample distortion during preparation and cut.

In order to ensure a tight and uniform clamping of the sample, a calibrated 1 mm layer of neoprene rubber is intercepted between the sample and the clamping device as per Fig. 9, which also provides details of the new clamping assembly. The cut test is then performed as per the standard ASTM operating procedure.

This method also leads to a much lower cut variability, CV % = 18, for a blade displacement of 10 mm under a load of 800 g. The normal probability plot is linear, and the Anderson-Darling normality test provides a P-value of 0.427, which also supports normality. Fifty-six data points were gathered. Figure 10*d* details these results.

None of the previously described approaches were designed and conducted to be readily integrated in a norm procedure; nonetheless both support the fact that the variability of the ASTM blade cut distance measurement can be improved by a twofold ratio. This is performed via an improved sample holding and limitations of the fiber freedom to move during testing. As mentioned before, the interpenetration of a resin in a textile knitted or woven fabric tends to reduce its cutting performance; therefore the coating approach suggested can only be used for comparative purpose.

Conclusion

Several test methods based on their equivalent standard procedures were reviewed beyond their normal application frames.

In term of harmonization, the considered test conditions reinforce the appropriateness of the convergence toward a single norm allowing for multiple choices at the equipment and procedure levels to best represent the suitability of the material versus the expected cut protection level. From this study one can appreciate as an improvement of the EN norm the possibility to test as per ASTM or ISO norm procedure when abrasive materials beyond level 3 are tested. Similarly, the interchangeability of the ASTM and the ISO apparatus appears more and more useful as the fundamental understanding of test condition limits increases. Several situations studied in the present work support this aspect.

In terms of evolution, the emergence of high cut performance materials reinforced with glass or steel fibers, the market progression of coated protective equipments, the improvement of the test variability, and the substitution of the calibration material seem to be good focus points for reflection in the near term. Several conclusions drawn in the course of this study may serve as a basis. The dulling effect of glass fibers toward the blade cutting edge can be misleading in terms of cut performance analysis. Similarly, the artificial cut contribution of polymeric dots deposited on the textile substrate can also cause difficulties in terms of the cut performance interpretation. Those two specific situations are avoided when the blade cutting edge is in contact with the sample only once during the testing. ASTM or ISO are therefore more favorable in such cases.

A large amount of work remains to be done to fully integrate all the components of the identified improvement in a global understanding of how the cut mechanism and testing can evolve to be better predictive tools.

Acknowledgments

The authors wish to express their gratitude to Imre Horvath for his continuous support and comments during the various reviews of this work, to André Miret-Casas for his meticulous technical assistance and enthusiasm, and to Gerry O'Donnell for the independent evaluation and analysis of the materials tested.

References

[1] BS EN388, 1994, Protective Gloves against Mechanical Risks.

[2] ASTM Standard F 1790-97, "Standard Test Method for Measuring Cut Resistance of Materials Used in Protective Clothing," ASTM International, West Conshohocken, PA.

[3] ISO13997, ISO / TC 94/SC 13/WG 5 N 173, Protective Clothing- Mechanical Properties- Determination of Resistance to Cutting by Sharp Objects.

[4] Lara, J., Turcot, D., Daigle, R, and Payot, F., "Comparison of Two Methods to Evaluate the Resistance of Protective Gloves to Cutting by Sharp Blades," *Performance of Protective Clothing: Fifth Volume, ASTM STP 1237*, James S. Johnson and S. Z. Mansdorf, Eds., ASTM International, West Conshohocken, PA, 1996.

[5] Payot F., "Measurement and Control Method for Cutting Resistance of Protective Gloves," *Performance of Protective Clothing; Fourth Volume, ASTM STP 1133*, James P. McBriarty and Norman W. Henry, Eds., ASTM International, West Conshohocken, PA, 1992.

[6] Massé, S., Lara, J., Sirard, C., and Daigle, R., "Basic Principles Used in the Development of a New Cut-Test Apparatus for Standardization," *Performance of Protective Clothing: Sixth Volume, ASTM STP 1273*, Jeffrey O. Stull and Arthur D. Schwope, Eds., ASTM International, West Conshohocken, PA, 1997.

[7] Lara, J., Turcot, D., Daigle, R., and Boutin, J., "A New Test Method to Evaluate the Cut Resistance of Glove Materials," *Performance of Protective Clothing: Fifth Volume, ASTM STP 1237*, James S. Johnson and S. Z. Mansdorf, Eds., ASTM International, West Conshohocken, PA, 1996.

[8] Tejani, N., Blocker, R., Schiffelbein, P., and Rivet, E., "Cut Protection Performance Test for Measuring Cut Resistance of Materials Used in Protective Clothing," *Performance of Protective Clothing: Sixth Volume, ASTM STP 1273*, Jeffrey O. Stull and Arthur D. Schwope, Eds., AASTM International, West Conshohocken, PA, 1997.

[9] American National Standard for Hand Protection Selection Criteria, ANSI/ISEA 105-2000.

Journal of ASTM International, May 2005, Vol. 2, No. 5
Paper ID JAI12148
Available online at www.astm.org

T. B. N. Vu,[1] *T. Vu-Khanh,*[1] *and J. Lara*[2]

Progress in the Characterization of the Cutting Resistance of Protective Materials

ABSTRACT: A sliding sharp edge that penetrates a material is one of the most dangerous cases of cutting because it requires the smallest applied normal load. This study aims to analyze the cutting mechanics and mechanism of protective rubber materials in the presence of friction and the effect of the material's mechanical behavior and the blade's sliding velocity on the material's cut resistance. The International Standard ISO 13997 cut test method, which consists of measuring the distance that a straight blade slides horizontally to cut through a material under a constant applied normal force, was used to investigate the cutting phenomena. In practice, the cut resistance of a material is given by the contribution of the material's intrinsic strength and the frictional distribution between the material and the blade that slides and penetrates it. This study demonstrates that two types of friction are involved in material cutting: a macroscopic friction induced by the gripping of the material and by the applied normal load on the two sides of the blade; and a sliding friction associated with cut-through of the material that occurs along the face of the blade tip. For rubber materials, commonly used in protective gloves, the adhesion force due to the gripping of the material on the blade edge could be several times greater than the friction due to the applied normal force. Thus, the cutting energy required to break the molecular chains in rubber materials is much smaller than the energy dissipated by friction. For these materials, the elastic modulus, the structure of the material, as well as the sliding velocity, have a significant effect on the friction. Therefore, all of these properties can affect the cutting resistance results. A better understanding of the cutting mechanism in protective materials is a fundamental step in developing better performing protective materials.

KEYWORDS: cut test method, cut resistance, protective glove, glove material, rubber material

Introduction

In the standard test method ISO 13997, the cut resistance of a material is expressed as the cutting force that must be applied on a straight blade to cut through the material in a 20-mm blade stroke [1]. The material cut resistance evaluated by this test method depends on the geometry of the sample holder, the thickness of the specimen, and the friction forces that develop when the sliding blade penetrates the material. It is known that a material's cut resistance is given by the contribution of the intrinsic strength of the material and the frictional distribution between the material and the blade that slides and penetrates it [2]. There are a few works reported on cutting in the absence of friction to characterize the intrinsic strength of rubber materials [3–5]. However, under practical conditions of use, the contribution of friction to the cut resistance of rubber materials in protective gloves is significant. The higher the friction forces involved in material cutting, the higher the material cut resistance will be [2].

Manuscript received 19 December 2003; accepted for publication 4 October 2004; published May 2005. Presented at ASTM Symposium on Performance of Protective Clothing: Global Needs and Emerging Markets: 8th Symposium on 13 January 2004 in Tampa, FL.

[1] Université du Québec – École de technologie supérieure, Département de génie mécanique, 1100, rue Notre-Dame Ouest, Montreal, Québec, Canada H3C 1K3.
[2] Institut de recherche Robert Sauvé en santé et en sécurité du travail, 505, boul. de Maisonneuve Ouest, Montréal, Québec, Canada H3A 3C2.

Analysis of Cutting Phenomena with the Standard Cutting Test Method

When a blade slides and penetrates a rubber specimen under an applied normal load, two types of contacts occur simultaneously. One is contact at the blade tip, which causes cutting. The shear stress, τ, at the blade-material contact layer has to overcome the molecular cohesion and the hysteresis friction arising from the viscoelastic behavior of the material. The other is an elastic contact of material on both sides of the blade, which is due to the lateral compression arising from the sliding blade that penetrates the elastic material and the pushing due to the applied normal load.

From the above analysis, it is clear that two aspects of friction are involved. One is a macroscopic friction for the entire system, induced by material compression on the two sides of the sliding blade and by the applied normal load. The other is a sliding friction associated with the material cut-through that occurs along the face of the blade tip. The pressure due to the applied normal force is very high in the vicinity of the tip [6–8]. However, the diameter of the blade tip is very small, about 1–2 μm, resulting in a negligible friction force. It is known that the energy dissipated as lateral friction is much higher than the intrinsic energy required for cutting through a material [1].

Theory Review

The classical law of Amontons-Coulomb on sliding friction states that the friction force is proportional to the externally applied load and independent of the apparent contact area. On a microscopic scale, sliding friction force is commonly related to contributions of adhesion and hysteresis. Adhesion friction is due to the formation and breakage of adhesive junctions, which are the results of molecular interaction on the contact surfaces. Hysteresis friction is the internal friction of a material, which arises from the dissipation of energy during cyclic deformations of microscopic protuberances on surfaces and depends on the rubber's viscoelastic properties. Rubbers with very low modulus and large elastic deformation are soft enough to result in intimate contact with hard, rough surfaces. During sliding, the molecular chains of rubber form local junctions with the substrate, and these bonds stretch, then break and relax before forming new junctions elsewhere [9]. At low speed, the surface area of a rubber material in contact with a substrate will deform in such a manner as to completely follow the short wavelength surface roughness profile [10–11]. When the substrate is very smooth and clean, the mechanism for friction is solely molecular adhesion at the interface [10–12]. If the temperature remains constant, the velocity dependence of the friction force must be attributed to a rate-dependent effective strength of the adhesive bonds [7,12]. The Johnson-Kendall–Roberts (JKR) theory [13] showed that even if the normal load is equal to zero, an adherence force exists at the contact area of soft rubber and hard surfaces.

Generally, the total friction force due to adhesion, F_{ah}, can be written as follows [14]:

$$F_{ah,} = \sum_{i=1}^{M} A_i \tau_i \tag{1}$$

where A_i is the local macroscopic area of contact between an elastomer and a rough rigid substrate, and τ_i is the local effective shear strength of the interface.

Ludema and Tabor [15] assumed that the adhesion bonds at the interface are as strong as the rubber itself. They suggested that tensile stress at break σ_c is a measure of shear strength τ at the contact surface. In the proposed theory, τ has been assumed to have the same speed dependence as the tensile stress σ_c at the point of failure. Based on the Voigt model, Moore [14] improved this simple theory and suggested an equation for shear strength as follows:

$$\tau = \sqrt{E'^2 + (\omega\eta)^2}\ e = E'\ \sqrt{\left[1 + (\tan\delta)^2\right]}e \qquad (2)$$

where e represents the strain in the thin surface layer, E' is elastic modulus, ω is frequency, η is viscosity, and $\tan\delta = \eta\,\omega / E'$.

This study aims to analyze the cutting mechanics and mechanism in the presence of friction, and the influence of mechanical behaviors of materials and the sliding velocity on the cut resistance of rubbers. The International Standard ISO 13997 test method is used as a reference to investigate the cut resistance of some selected protective materials.

Experimental

Cut Test Apparatus

An instrumented version of the cut test apparatus TDM 100[3] described in the ISO 13997 standard was used for this study and is shown in Fig. 1*a*. The test apparatus was equipped with a Vexta step motor, model PK 264-03 from Oriental Motor U.S.A. Corporation, which can run at speeds from 1–900 mm/min. The motor is turning at 200 steps per revolution, which correspond to a blade displacement of 1.0 mm, allowing the blade displacement to be measured with a precision of 0.005 mm.

A load cell of 44 N (10 lb) capacity was fixed at one extremity of the blade-holder and allowed the horizontal friction force to be measured and is represented in Fig. 1*b*. The blade was GRU-GRU type, model 88-0121, from American Razor Co[4], and was made of high carbon steel, which was heat-treated to a hardness of 78–80 R30N. The blade was 70 mm long, 18 mm wide, and 1.0 ± 0.05 mm thick, had a bevel width of 2.5 ± 0.2 mm along the straight edge, the apex angle 2α was from 32° to 36°, and the angle, $2\alpha_1$, of two sharp edge sides was 17°. Figure 2*a* is the optical micrograph of the surface roughness of the blade edge magnified 200 times, and Fig. 2*b* illustrates the blade geometry.

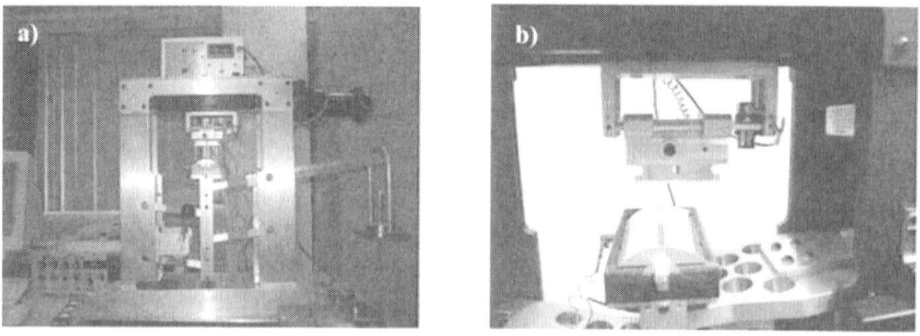

FIG. 1—*Pictures of the instrumented version of the TDM-100 (a) and the blade holder with an attached load cell to measure the horizontal friction force (b).*

[3] RGI Industrial Products, Inc., 755 Pierre Caisse, St-Jean-sur-Richelieu, Quebec, Canada J3B 7Y5.

[4] American Safety Razor Co., Razor Blade Lane, Verona, VA 24482, U.S.A.

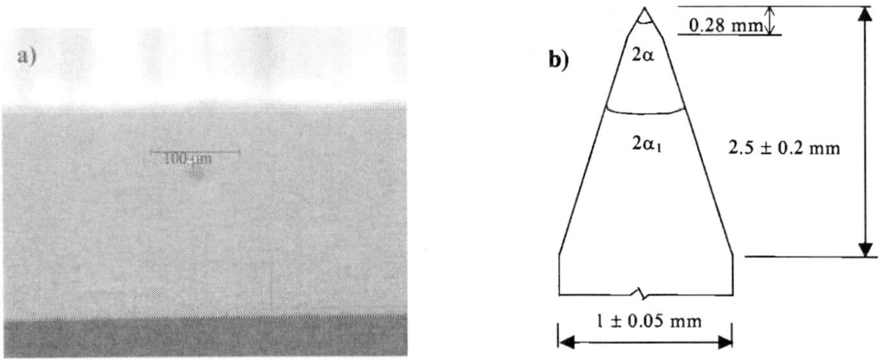

FIG. 2—*Picture of the blade edge magnified 200 times (*a*) and schematic representation of the blade geometry (*b*).*

The cut test was performed by applying a constant force at the blade-material point of contact, perpendicular to the sample. The straight blade moved horizontally during the test with a controlled constant speed. The specimen was attached with double-sided adhesive tape on the curved surface (which had a radius of 38 ± 0.5 mm) of the sample holder. Directly under the specimen, a narrow piece of aluminum foil 8 mm wide with a thickness smaller than 0.03 mm was installed at the center of the tape. When the specimen was cut through, the blade came in contact with the electric conductive foil, which caused the electronic circuit to stop the test automatically. The load cell installed on the side of the blade-holder measured the horizontal friction force developed during the test. A control unit, which controlled the speed and stopped the experiment as well as a computer for data acquisition and treatment, completed the system. For this study, most of the cutting tests were performed at 150 mm/min. In order to characterize the effect of blade speed on the cutting results, the tests were performed at different blade velocities between 1 and 900 mm/min.

Materials

Various materials with different thicknesses were used for this study: Neoprene sheets with thicknesses of 0.41 mm, 0.79 mm, and 1.57 mm from Fairprene Industrial Products Co. USA; Nitrile Solvex gloves, 37-165 from Ansell, made in England; and natural rubber gloves, model 574 form Kachele-Cama latex (KCL) GmbH, D-36124 Elchenzell of Strantfords Ltd., Germany. Also, a glass fiber fabric, 8-harness satin (aircraft cloth 38 in.') supplied by Progress Plastiques Canada was used as an example of a material with a very low friction coefficient.

Results and Discussion

Effect of Gripping Force on Cutting

When a blade slides horizontally and penetrates a material, a horizontal force due to the friction between the blade and the material appears. This friction force results from the gripping of the elastic material and the applied normal force. In order to investigate the contribution to the friction force due to gripping, tests were carried out on specimens of neoprene 1.57 mm thick with various applied loads. The blade slid at a constant speed of 150 mm/min to penetrate the specimen. When the blade reached the lower surface of the specimen, the electric stopping circuit was disconnected. At that time, the end of the lever arm where the weights were installed touched a support that allowed the vertical movement of the specimen to be blocked. At that moment, the normal force applied on the sample became zero, while the blade continued to slide through the material cross–section. An example of the results obtained for this kind of test is shown in Fig. 3, where the friction forces versus blade displacement obtained with an applied load of 9 N are represented.

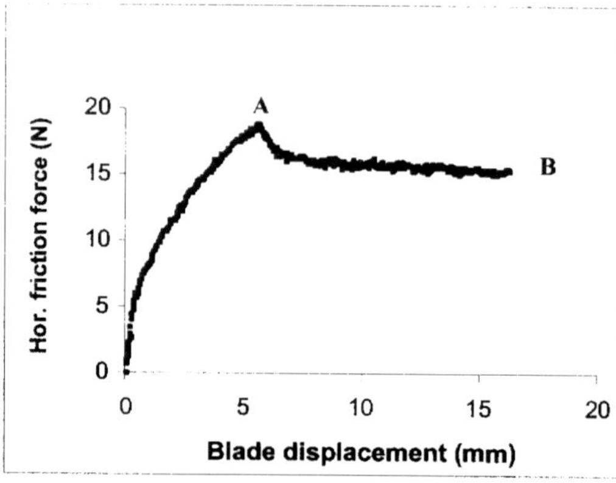

FIG. 3—*Horizontal friction forces in cutting neoprene 1.57 mm thick (F_N = 9N, curve 0-A) and in the blade sliding through the material cross-section (F_N = 0, curve A-B).*

When the blade penetrates the material, the horizontal friction force increases gradually with the sliding distance to reach a maximum value corresponding to the moment of full penetration of the blade (illustrated by curve 0-A in Fig. 3). The signal recorded after the maximum force at point A on the figure represents the frictional force due to gripping, since cutting had been completed (illustrated by curve A-B in Fig. 3). It can be seen that during the time that the blade slid without cutting, the contact area was unchanged, and the friction force remained nearly constant. Given that the blade surface is relatively smooth, the small friction force due to hysteresis was negligible; the friction force due to gripping mostly resulted from the molecular adhesion of the material-blade contact surfaces. The high value for the friction force obtained in these experiments showed that there was strong adhesion between the material cross-section and the blade. Furthermore, because the friction force remained high while the blade continued to slide and no normal force was applied suggests that the molecular junctions created between the blade and the rubber material may not be broken by the release of the elastic stress. Hence, the contact area due to material gripping when the blade cut through the specimen (at point A) and the contact area at any time when the blade slid without cutting (curve A-B) may be nearly identical. We assumed that the value at which the average friction force becomes constant corresponds to material gripping and is represented by F_g. The maximum friction force F_{hf} at point A results from the friction force F_{fl} due to the applied normal load and the friction force F_g due to material gripping. Table 1 and Fig. 4 present the friction contributions of the normal force and the gripping force that were calculated from the tests performed on neoprene 1.57 mm thick under various applied normal loads, according to Eq 3:

$$F_{hf} = F_g + F_{fl} \quad (3)$$

TABLE 1—*Calculated contributions of gripping and normal forces on cutting tests on neoprene 1.57 mm thick.*

Applied Load (N)	Max. Horizontal Force (N)	Friction Force Due to Gripping F_g (N)	Friction Force Due to Applied Normal Load F_{fl} (N)
5	16.90	13.33	3.57
7	17.13	12.97	4.16
9	18.40	13.40	5.00

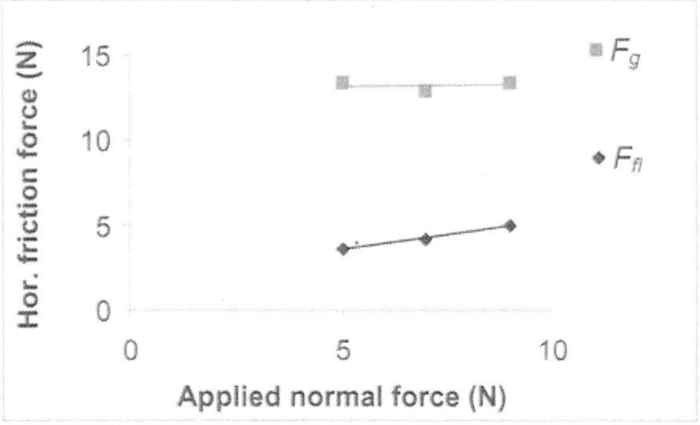

FIG. 4—*Calculated values for friction forces due to material gripping and normal force in cutting through neoprene 1.57 mm thick.*

Figure 4 shows that for the different applied loads, friction forces due to gripping are the same and are independent of the applied normal load, while the friction forces arising from the applied load are proportional to the applied load. Therefore, if the friction forces due to the blade tip are neglected, the calculated values for the horizontal frictional force due to material gripping are many times larger than the horizontal frictional force due to the applied normal load. This clearly indicates that the energy expended in cutting through with a sliding movement depends mostly on the interfacial adhesion of the material and the blade sides.

Friction Mechanism on Cutting

Figures 5 and 6 illustrate the relationship between the applied normal load and the horizontal friction forces on cutting various thicknesses of neoprene, nitrile, natural rubber, and glass fiber fabric. The horizontal friction force is measured when the specimen has just been cut through. As previously mentioned, the horizontal friction force is due to the applied normal force and the gripping of the rubber material on the blade edges.

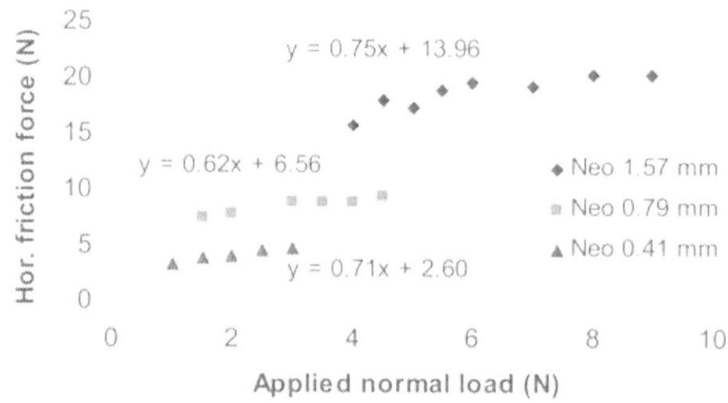

FIG. 5—*Horizontal friction forces versus applied normal loads in cutting tests for different thicknesses of neoprene.*

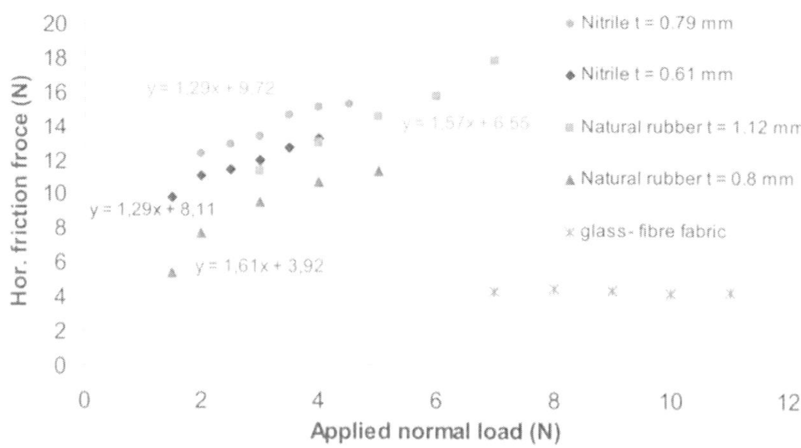

FIG. 6—*Horizontal friction force as a function of the applied normal loads in cutting tests on two different rubber glove materials and a glass fiber fabric.*

The results show that horizontal friction forces and applied normal loads have a linear relationship, as represented by

$$F_{hf} = \mu_n F_N + F_{adh} \tag{4}$$

Where F_{hf} is the horizontal friction force, F_N is the applied normal load, and μ_n (which represents the slope of the curve) is the ratio of the horizontal friction force over the applied normal load. The second term F_{adh} is a constant, which is independent of the applied load and related to the contact area A_c, and consequently, is related to material gripping. The results represented in Figs. 5 and 6 show that μ_n is a constant for the same type of material and is independent of material thickness. Table 2 represents the calculated values of the adhesion force F_{adh}, using Eq 4 and illustrated in Figs. 5 and 6, and the gripping forces per unit of contact area. The value of gripping force F_g from Table 1 and the value of the term F_{adh} in Table 2 for neoprene 1.57 mm thick are nearly the same. This confirms that the term F_{adh} is related to the gripping of the material on the blade and is proportional to the contact area.

TABLE 2—*Adhesion force F_{adh} in cutting various thicknesses of neoprene, nitrile, and natural rubber.*

Material	Contact Area A_c (mm^2)	Adhesion Force due to Friction Between Blade Sides and Material F_{adh} (N)	F_{adh} / A_c (N/mm^2)
Neoprene 1.57 mm	47.34	13.96	0.29
Neoprene 0.79 mm	16.77	6.15	0.37
Neoprene 0.41 mm	6.24	2.60	0.41
Nitrile 0.79 mm	16.77	9.72	0.58
Nitrile 0.61 mm	11.36	8.11	0.71
Natural rubber 1.12 mm	28.40	6.56	0.23
Natural rubber 0.80 mm	17.10	3.92	0.23

When a blade cuts a soft elastic material, the two surfaces of the cross-section of the material will strongly squeeze and adopt the corrugations of the blade, which have an average wavelength λ of roughness in a range from 100–1000°A. Rubber particles from cutting and wear may fill most of the cavities of the blade bevels. If the blade slides at a speed from 1–900 mm/min, the frequencies of the asperities of the rubber surface, $\omega = v/\lambda$, are in the range of 10^2 to 10^6 s^{-1}. This suggests that the real contact area of the blade and the material is nearly equal to the apparent contact area and that the adhesion component may be much higher than the hysteresis distribution. At a distance of 0.28 mm from the blade tip (see Fig. 2*b*), the blade surface is rougher than the upper part. Pressure is concentrated in this zone, and therefore a large part of the friction force may occur here. For this reason, the values of F_{adh} / A_c for thinner specimens are slightly higher, as can be seen in Table 2.

Considering the resultant equilibrium forces in the vertical direction at any time in cutting (see Fig. 7), we obtain:

$$F_N - \mu_1 F_{load} \cos\alpha - F_{load} \sin\alpha = 0 \quad (5)$$

Where μ_1 is the blade-material friction coefficient due to vertical movement; F_{load} is the resultant of the pressure on the material, which is due to the applied normal load; F_N is the applied normal load; and α is half of the apex angle of the blade (see Fig. 7):

Hence

$$F_{load} = \frac{F_N}{\mu_1 \cos\alpha + \sin\alpha} \quad (6)$$

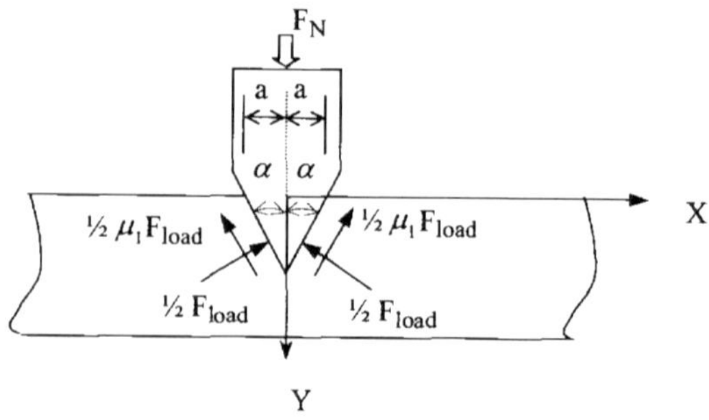

FIG. 7—*Diagram of distribution of the forces due to applied normal load.*

The contribution of the horizontal friction force, which is caused by the normal force, has the value of:

$$F_{fl} = \mu_n \, F_{load} = \mu_n \frac{F_N}{\mu_1 \cos\alpha + \sin\alpha} \tag{7}$$

If,

$$\mu_n \, F_{load} \approx \mu_n \, F_N \tag{8}$$

Hence,

$$\mu_1 \approx (1 - \sin\alpha)/\cos\alpha \tag{9}$$

According to Eq 9, the friction coefficient in the vertical direction is independent of pressure distribution and only dependent on the apex angle of the blade.

The theory of elasticity [8] proved that in the frictionless condition, the mean pressure at the contact surface of a symmetric wedge is independent of load applied and only dependent on the apex angle of the wedge. Experimenting with sliding the cones of various semi-apex angles α on a lubricated rubber, Moore [14] showed that sliding friction coefficient μ depends only on the semi-apex angle and not on the pressure distribution. And heavy tearing occurred when α is approximately 30°.

If a semi-angle of the apex of the blade α is from 16–18°, Eq 9 gives $\mu_1 \approx 0.70$–0.72. Table 3 represents the values of the horizontal friction force $\mu_n F_{load}$ calculated using Eq 13 with $\mu_1 = 0.70$, $\mu_n = 0.70$ for neoprene, $\alpha = 18°$; and the experimental values of the friction force due to the applied load on neoprene 1.57 mm thick. The results show that the theoretical calculations nearly agree with the experimental values.

In cutting rubber, there is a strong adhesion force at the interfaces, which is the main cause of the gripping friction force F_{adh}. This friction force will be proportional to the contact area A_c and the shear strength τ of the molecular junctions formed at the contact surface.

Hence, the gripping force F_{adh} can be written:

$$F_{adh} = \tau \, A_c \tag{10}$$

TABLE 3—*Comparing the theoretical values and experimental values of the horizontal friction force resulting from applied normal loads on neoprene 1.57 mm thick.*

Applied Normal Load (N)	μ_n F_{load} (N) (see Eq 13)	Experimental Results of Horizontal Force Due to Applied Normal Load	
		μ_n F_N (N) (see Fig. 5)	From Gripping Test (see Table 1) (N)
5.0	3.60	0.7*5 = 3.50	3.57
7.0	5.00	0.7*7 = 4.90	4..17
9.0	6.50	0.7*9 = 6.30	5.00

Moore's equation (see Eq 2) can be used to calculate shear strength. If the blade's sliding speed changes from 1–900 mm/min, the frequencies of the asperities of the rubber surface, $\omega = v/\lambda$, are in the range of 10^2 to 10^6 s^{-1}. For a typical rubber at room temperature, with elastic modulus E (ω) $\approx$E (0) for $\omega < \omega_c = 10^5$ s^{-1} [16], friction due to internal deformation may be neglected, and Eq 2 becomes:

$$\tau \approx E' e \qquad (11)$$

Elongation to break is about 340 % for neoprene, about 400 % for nitrile, and more than 800 % for natural rubber. During sliding, the rubber asperities may be stretched some distance. Equation 11 can be used to calculate the strain necessary to break molecular junctions at the material-blade interface. The tensile modulus is 4.56 MPa for neoprene, 7.39 MPa for nitrile, and 1.70 MPa for natural rubber. These values were calculated from tensile tests with dumbbell samples from the ASTM standard at 8 % elongation, which were measured by laser extensometer. The test speed is the same speed as used for cutting, namely 150 mm/min. Supposing that perfect blade-material contact occurs, Table 4 gives the experimental values of the mean strain to break the molecular bonds at the contact surface.

TABLE 4—*Value of the mean elongation to break of adhesion bonds.*

Material	A_c (mm^2)	Gripping Force F_{adh} (N) (see Figs. 5-6)	Gripping Force/A_c (N/mm^2)	E (Mpa)	e (%)
Neoprene 1.57 mm	47.34	13.96	0.30	4.56	6.58
Neoprene 0.79 mm	16.75	6.15	0.37	4.56	8.11
Neoprene 0.41 mm	6.24	2.60	0.41	4.56	8.99
Nitrile 0.79 mm	16.75	9.72	0.57	7.39	7.71
Nitrile 0.61 mm	11.36	8.11	0.71	7.39	9.61
Natural Rubber 1.12 mm	28.40	6.55	0.23	1.70	13.53
Natural Rubber 0.80 mm	17.10	3.92	0.23	1.70	13.53

Nitrile and neoprene are filled rubbers. The elongation to break the junctions at the interfaces of neoprene and nitrile is about 8 % and about 14 % for unfilled natural rubber. It known that an 8 % strain amplitude is large enough to completely break down the filler network of rubbers [16].

Surprisingly enough, the value of elongation, *e*, can be rewritten as $K/2\pi$, and the resultant horizontal friction force arising from the sliding blade that penetrates a rubber specimen can be approximately represented by:

$$F_{hf} = \mu_n F_N + E\, e\, A_c \; = \mu_n F_N + \mathrm{K}\frac{E}{2\pi}\, A_c \tag{12}$$

Where K is nearly equal to 0.5 for filled rubbers and 0.8 for natural rubber. Equation 12 can be verified by the cutting test with glass fiber satin fabric and high pre-strained rubber samples. Table 5 gives the values of the adhesion forces, which are calculated by using Eq 12, and from the relationship between horizontal friction forces and the applied normal loads.

During the cutting of glass fiber satin fabric (see Fig. 6), the applied load concentrated almost at the blade tip, and pressure on the two sides of the blade was nearly zero, resulting in the first term $\mu_n F_N$ of Eq 12 being nearly equal to zero. On the other hand, when cutting tests were performed on pre-strained samples of neoprene 1.57 mm thick (see Fig. 8), due to the high pre-strain when the material has just been cut, it retracted immediately. The friction arising from adhesion has been eliminated, so that the second term $\mathrm{K}\frac{E}{2\pi} A_c$ became nearly equal to zero. The effect of pre-strain on cut resistance will be presented in another paper.

Table 5. Theoretical and experimental values of adhesion force of different rubbers

Material	A_c (mm^2)	K	Theoretical Values (Acc. to Eq 12): $\tau = \mathrm{K}\frac{E}{2\pi}$ (N/mm^2)	Theoretical Values (Acc. to Eq 12): F_{adh} (N)	Experimental Results: F_{adh} (see Figs. 5–6) (N)	Experimental Results: F_{adh} / A_c (N/mm^2)
Neoprene 1.57 mm	47.34	0.50	0.36	17.18	13.96	0.30
Neoprene 0.79 mm	16.75	0.50	0.36	6.08	6.15	0.37
Neoprene 0.41 mm	6.24	0.50	0.36	2.26	2.60	0.41
Nitrile 0.79 mm	16.75	0.50	0.70	11.82	9.72	0.57
Nitrile 0.61 mm	11.36	0.50	0.70	8.02	8.11	0.71
Natural Rubber 1.12 mm	28.40	0.80	0.22	6.15	6.55	0.23
Natural Rubber 0.80 mm	17.10	0.80	0.22	3.63	3.92	0.23

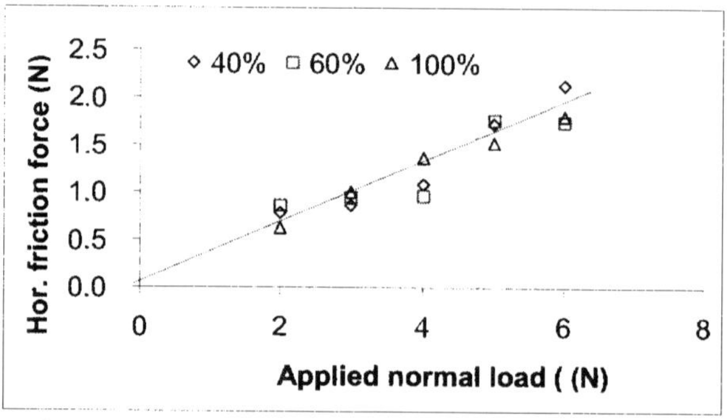

FIG. 8—*Effect of pre-strain on horizontal friction force of neoprene 1.57 mm thick.*

Effect of the tensile modulus of rubbers on blade displacement and the horizontal friction force was verified in cutting aging neoprene. The strength of rubbers can be seriously affected by oxidation, especially in the high temperature. During thermal aging, main chain scission, crosslink formation, and crosslink breakage may occur. Neoprene gradually became brittle. As seen in Fig. 9, there was a substantial increase in the tensile modulus. Figure 10 shows that the tensile modulus increased with aging time, resulting in an increase in horizontal friction force. Consequently, as seen in Fig. 11, blade displacement increased.

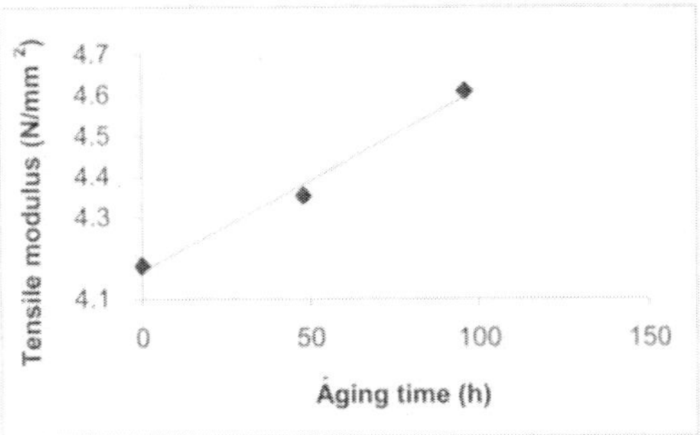

FIG. 9—*Tensile modulus at 8 % elongation versus aging time of neoprene 1.57 mm thick.*

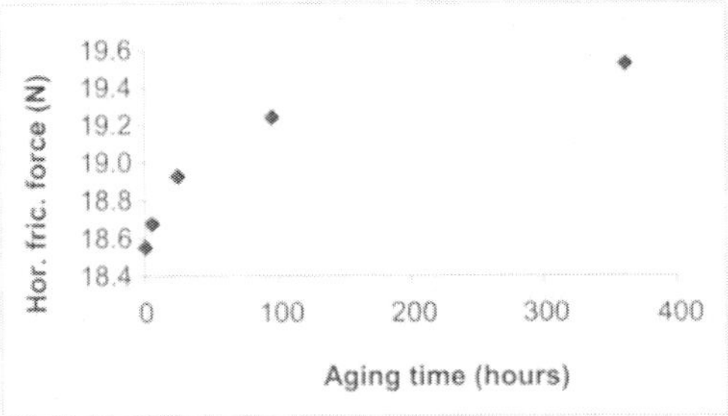

FIG. 10—*Effect of heat aging at 100°C on maximum horizontal force of neoprene 1.57 mm thick, cutting with the applied load of 5N.*

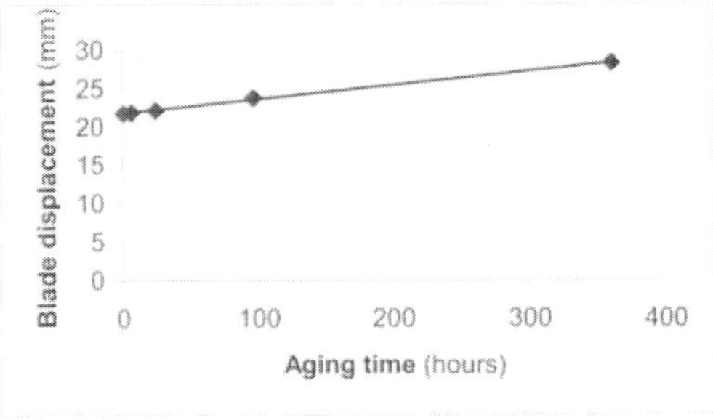

FIG. 11—*Effect of heat aging at 100°C on cut resistance of neoprene 1.57 mm thick, cutting with the applied load of 5N.*

Effect of Sliding Speed on Cutting

When investigating the viscoelastic effect of elastomers, Cho & Lee [5] showed that the cutting energy in the absence of friction is almost constant at a low effective cutting rate (crosshead speed < 10^{-5} m/s), which is the threshold condition in the cutting process. The cutting energy increases abruptly as the effective cutting rate increased and became nearly constant at a high effective cutting rate. This result also agreed with cutting in the presence of friction using the ISO 13997 standard cutting test method. The results in Figs. 12 and 13 illustrated the effect of sliding speed on cutting energy and blade displacement for nitrile 0.61 mm thick, natural rubber 0.84 mm thick, and neoprene 1.57 mm thick. The applied load was 2N to cut through nitrile and natural rubber, and 5N for neoprene. At room temperature, when the blade sliding speed was in the range of 1–100 mm/min, the blade displacement, the horizontal friction force, and the cutting energy increased abruptly. However, they increased very slightly when the blade sliding speed was in the range of 100–900 mm/min.

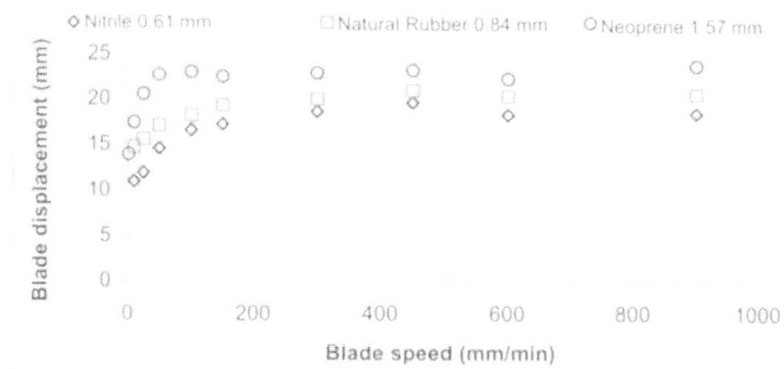

FIG. 12—*Effect of sliding velocity on blade displacement for nitrile, natural rubber, and neoprene.*

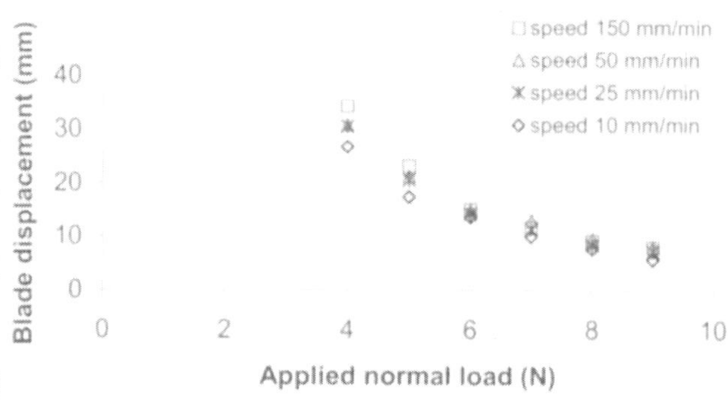

FIG. 13—*Blade displacement versus applied normal load at different speeds in cutting of neoprene 1.57 mm thick.*

It can be seen in Fig. 12 that the cut resistance results for rubber materials increases more than 1.5 times if the blade sliding speeds change from 1 to 100 mm/min.

Figures 13 and 14 show the relationship between blade displacement and the maximum horizontal friction force with an applied normal load on neoprene 1.57 mm thick for different speeds. At low speed, the effect on friction due to the external load is very small, and the values of friction coefficient μ_n due to the applied load are nearly the same, while the horizontal friction force F_{adh} due to surface adhesion increased significantly. These results agree with the conclusion of K. A. Grosch [12] about friction of rubber at very low velocities. The magnitude of F_{adh} depends on the number of adhesive bonds, and the effective strength of the adhesive bonds increases with the increase of velocity.

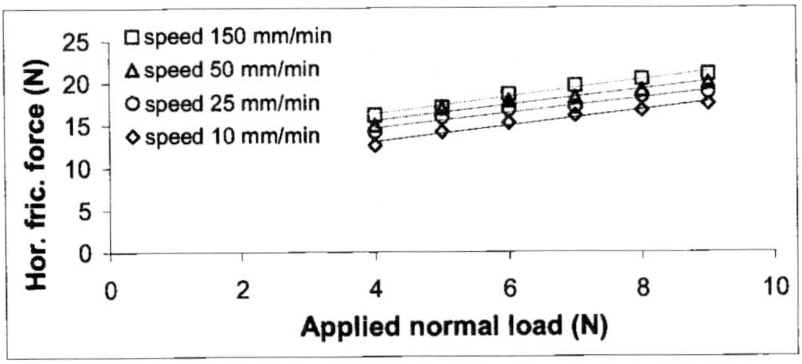

FIG. 14—*Horizontal friction force versus applied normal load at different sliding speeds. Test is on neoprene 1.57 mm thick.*

Conclusions

Cutting due to the contribution of both a normal force and the sliding of a sharp object more exactly reflects the cut behavior of the protective material being used. The energy required to break molecular chains of rubber is very small, about $5GJ/m^3$ to break C-C chains [4,5], while a large part of the energy used for cutting is dissipated by friction, as demonstrated in the present study. Therefore, an increase in friction can enhance the cut resistance of materials. Using the ISO cutting test method, it has been shown that the horizontal friction force arising from blade penetration into a rubber material is generated by the applied normal load and the material-blade interfacial adhesion. In most situations, the friction force arising from interfacial adhesion is much higher than the friction forces caused by the applied normal load. In the range of speeds from 1 to 900 mm/min and at light applied loads, the horizontal friction force as caused by the applied normal load was found to be proportional to the applied load and the proportional μ_n constant to be independent of the load applied and the material thickness. Molecular adhesion at the blade-material contact surface depends strongly on the material's tensile modulus. The strain-to-break of the adhesive bonds at the contact surface is about 8 % for filler rubbers and about 14 % for unfilled natural rubber. Under the same conditions, materials having higher tensile modulus will have higher cut resistance. Horizontal friction forces measured by the ISO cutting test method for thin sheets of rubber can be approximated by Eq 12.

Under the effect of thermal aging, the mechanical properties of neoprene are modified. Aged neoprene becomes harder, and the tensile modulus increases, resulting in an increase in the material cut resistance.

In the range of speeds from 1 to 100 mm/min, sliding speeds had a significant effect on adhesion force F_{adh}. This effect could be due to the increase of strength of the adhesive bonds with the speed. As mentioned above, blade displacement and horizontal friction force strongly depend on the adhesion force F_{adh}. Consequently, blade displacement and horizontal friction force increased with the increase of sliding speed. However, in the range of speeds from 100 to 900 mm/min, the adhesive force increased very slightly, resulting in nearly constant blade displacement and horizontal friction force.

References

[1] Lara, J., Turcot, D., Daigle, R., and Boutin, J., "A New Test Method to Evaluate the Cut Resistance of Glove Materials," *Performance of Protective Clothing, Improvement Through Innovation: Fifth Volume, ASTM STP 1237*, James S. Johnson and S. Z. Mansdorf, Eds., ASTM International, West Conshohocken, PA, 1994 pp. 23–32.

[2] Vu Thi, B. N., Vu-Khanh, T., Lara, J., "Effect of Friction on Cut Resistance of Polymers," *Journal of Thermoplastic Composite Materials*, accepted for publication.

[3] Lake, G. J. and Yeoh, O. H., "Measurement of Rubber Cutting Resistance in the Absence of Friction," *International Journal of Fracture*, Vol. 14, No. 5, 1978, pp. 509–526.

[4] Gent, A. N. and Wang, C., "Cutting Resistance of Polyethylene," *Journal of Polymer Science: Part B: Polymer Physics*, Vol. 34, 1996, pp. 2231–2237.

[5] Cho, K., Lee, D., "Viscoelastic Effects in Cutting of Elastomers by a Sharp Object," *Journal of Polymer Science: Part B: Polymer Physics*, Vol. 36, 1997, pp. 1283–1291.

[6] Kragelsky, I. V., Dobychin, M. N., and Kombalov, V.S., *Friction and Wear Calculation Methods*, Pergamon Press, 1982, pp. 56–57.

[7] Kummer, H. W. and Meyer, W. E., "Rubber and Tire Friction," *Engineering Research Bulletin B–80*, The Pennsylvania State University, 1960, pp. 42–43.

[8] Hill, D. A., Nowell, D., and Sackfield, A., *Mechanics of Elastic Contacts*, Butterworth–Heinemmann Ltd, 1993, pp.62–65.

[9] Schallamach, A., "The Velocity and Temperature Dependence of Rubber Friction," *Proc. Phys. Soc. B 66*, 1953, pp. 386–392.

[10] Persson, B. N. J., "Sliding Friction," *Surface Science Reports*, 33, 1999, pp. 83–119.

[11] Persson, B. N. J., *Sliding Friction – Physical Principles and Application*, Springer, Heidenberg, 2000, pp. 83–88.

[12] Grosch, K. A., "The Relation Between the Friction and Visco-Elastic Properties of Rubber," *Pro. Roy. Soc.*, ser. A, Vol. 274, No. 1356, 1963, pp. 21–39.

[13] Johnson, K. L., Kendall, K., and Roberts, A. D., "Surface Energy and the Contact of Elastic Solids," *Proceedings of the Royal Society of London*, A324, 1971, pp. 301–310.

[14] Moore, D. F., *The Friction and Lubrication of Elastomers*, Oxford Pergamon Press 1972, pp. 60–64, 220–222.

[15] Ludema, K. C. and Tabor, D., "The Friction and Visco–Elastic Properties of Polymeric Solids," *Wear*, Vol. 9, 1996, pp. 329–348.

[16] Persson, B. N. J., "Theory of Rubber Friction and Contact Mechanics," *J. Chem. Phys.*, Vol. 115, No. 8, 2001, pp. 3840–3861.

Journal of ASTM International, April 2005, Vol. 2, No. 4
Paper ID JAI12147
Available online at www.astm.org

C. Thang Nguyen,[1] *T. Vu-Khanh,*[1] *and J. Lara*[2]

A Study on the Puncture Resistance of Rubber Materials Used in Protective Clothing

ABSTRACT: The aim of this work is to investigate the intrinsic parameters that control the puncture resistance of materials used in protective clothing. The effect of various probe geometries and sample holder sizes on puncture tests results were investigated. Medical needles of different diameters and tip angles were also used as puncture probes. Various available commercial rubbers used in protective gloves were characterized. The findings of this study will be useful for recommendations for standard test method protocols to characterize the puncture resistance of protective clothing materials. When cylindrical probes with flat and rounded tip geometries are used, material puncture occurs when the strain reaches one material failure value independent of tip geometry. The failure strain is also independent of sample thickness, probe tip diameter, and the sample holder size. The results demonstrated that the tip angle of conical probes has a dramatic effect on puncture test results. The material puncture resistance obtained with conical probes with different tip angles having a tip diameter d_1 and a rod diameter d_2 can be interpolated from the corresponding values obtained with cylindrical probes having the same d_1 and d_2 diameters. On the other hand, cylindrical probes provide a more straightforward characterization of puncture and also eliminate the complex effect of conical probe-tip geometry. Test results with medical needles showed that the puncturing mechanism is different from that of conical and cylindrical probes. Puncture by medical needles involves cutting and fracture energy, whereas puncture by conical and cylindrical probes relates to failure strain.

KEYWORDS: puncture resistance, glove material, protective clothing, puncture probe

Introduction

Puncture resistance is among the major mechanical properties often required for protective clothing; however, the intrinsic material parameters controlling the puncture resistance of these materials are still unknown. The reported investigations are either qualitative and do not provide a fundamental understanding of the mechanisms controlling puncture [1–3], or they are only applicable to other classes of materials such as rubber blocks [4] and geomembranes [5–8].

At the present time, to evaluate the puncture resistance of protective clothing, standard tests ASTM F 1342 and ISO 13996 are the most commonly used methods. These tests have been developed to evaluate the puncture resistance of thin flexible materials. ASTM F 1342 [9] is designed for any type of protective clothing, including coated fabrics, laminates, textiles, plastics, elastomeric films, or flexible materials. This test method determines the puncture resistance of a material specimen by measuring the maximum force required for a conical puncture probe to penetrate through a specimen clamped between two plates with chamfered holes of diameter not less than 10 mm. However, this method is only a qualitative

Manuscript received 19 December 2003; accepted for publication 1 October 2004; published April 2005. Presented at ASTM Symposium on Performance of Protective Clothing: Global Needs and Emerging Markets: 8th Symposium on 13 January 2004 in Tampa, FL.
[1] Université du Québec – École de technologie supérieure, Département de génie mécanique, 1100, rue Notre-Dame Ouest, Montreal, Québec, Canada H3C 1K3.
[2] Institut de recherche Robert Sauvé en santé et en sécurité du travail, 505, boul. de Maisonneuve Ouest, Montréal, Québec, Canada H3A 3C2.

characterization and does not provide the intrinsic material properties that control puncture resistance.

Furthermore, it is clear that the puncture probe used in ASTM F 1342 (Fig. 1) is very different from actual pointed objects (some examples are shown in Fig. 2) and cannot accurately characterize puncture resistance to real objects. In fact, puncture of protective clothing materials by medical needles (Fig. 2*a*) has been studied [1–3], and the puncture resistances are found to be much smaller than those measured by ASTM F 1342. However, in these studies, the thickness is not taken into account, so that thicker samples give higher resistance, and the intrinsic material parameter controlling puncture resistance is still unknown. Therefore, additional studies should be carried out on puncture probes that are similar to real objects (e.g., medical needles can be used as puncture probes), and new testing methods should be developed for a more realistic characterization of the puncture resistance of protective clothing materials.

The aim of this work was to investigate the material parameters that control puncture resistance of thin rubber membranes. Various puncture probes alternative to the one in the ASTM standard F 1342 were analyzed to find the material parameters controlling puncture in this test. A puncture test using medical needles as puncture probes was also carried out. The puncture mechanisms, the effect of probe geometry, and the correlations with other mechanical properties were also investigated.

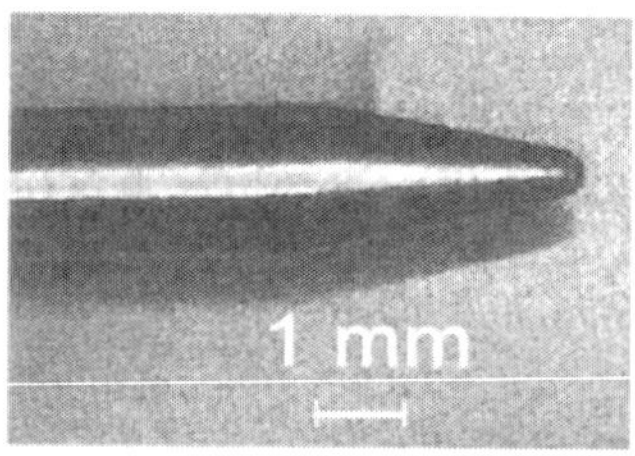

FIG. 1—*Puncture probe used in ASTM F 1342.*

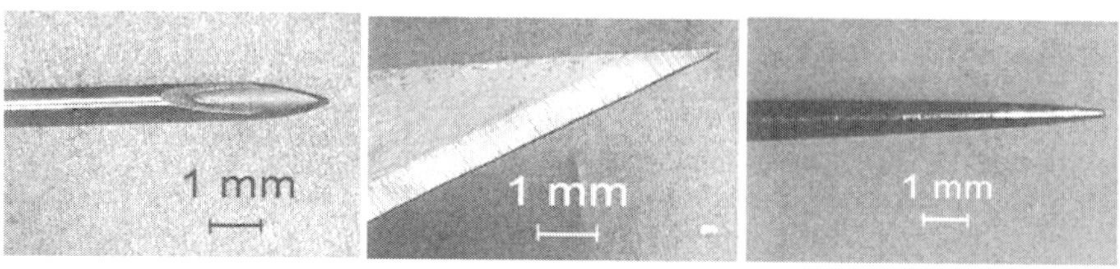

a) Medical needle b) Pointed tip of a knife c) Sewing needle

FIG. 2—*Actual pointed objects.*

Experimental

Materials

Three types of commercial rubbers commonly used for gloves were investigated. Neoprene sheets with three different thicknesses – 0.40, 0.78, and 1.57 mm – were obtained from Fairprene Industrial Products. The 0.30 mm thick nitrile samples were cut out from nitrile gloves manufactured by Ansell Co. The natural rubber samples 1.0 mm thick were cut out from NR Gloves, manufactured by Sandstrahler Co.

Puncture Test

The puncture test was carried out on an Instron 1137 universal-testing machine. The puncture probe was held by a pin chuck mounted on the load cell. The elastomer sample was clamped between two steel plates under pressurized air as shown in Fig. 3. The hole in the lower plate was chamfered to avoid stress concentration.

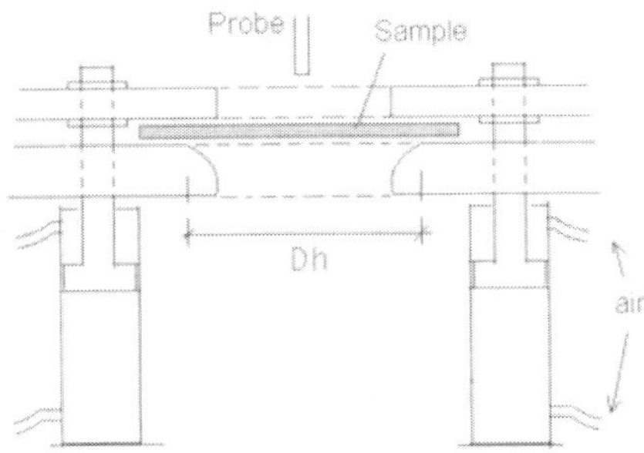

FIG. 3—*Sketch of sample holder.*

Three types of puncture probes were used: cylindrical probe (flat tip and rounded tip), conical probe (flat tip and rounded tip), and medical needle (Fig. 4). All probes were made of stainless steel.

For the cylindrical probes, the probe diameters of the flat tip (Fig. 4*a*) were 0.1 mm, 0.2 mm, 0.5 mm, 1.0 mm, 1.4 mm, and 2.5 mm. The diameters of the rounded tip (Fig. 4*b*) were 1.0 mm, 2.5 mm, and 5.0 mm.

For the conical probes with rounded tip (Fig. 4*c*), the conical extremities had an angle of 26°, and three probe sizes were available with rounded ends of 0.25 mm, 0.30 mm, and 0.50 mm radius. The first one corresponded to the probe used in ASTM F 1342. Conical probes with flat tip (Fig. 4*d*) and various tip angles were used to study the effects of tip angle on puncture resistance. Medical needles with various tip angles α, different diameters (0.2–1.0 mm), and hole diameters d' (Fig. 4*e*) were also used.

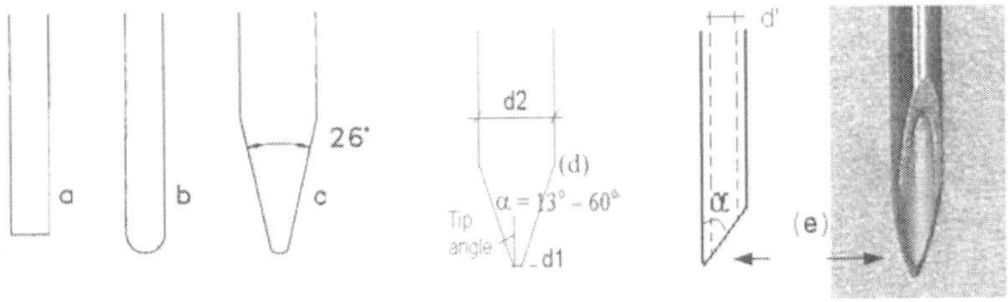

FIG. 4—*Puncture probes: cylindrical probes: flat tip* (a), *rounded tip* (b); *conical probes: rounded tip* (c), *flat tip* (d), *and medical needle* (e).

Tensile Test

The uniaxial tensile test was performed according to ASTM standard D 412 [10], using the C type dog-bone sample. The test was carried out at room temperature, with a crosshead speed of 100 mm/min. Elongation was measured by an MTS LX 500 laser extensometer.

Biaxial Test

Balloon tests [11,12] were carried out to determine the failure strain and stress in an equibiaxial state (Fig. 5). The pressure was measured and controlled by a pressure gauge. The maximum diameter of the balloon before burst was recorded by a MTS LX 500 laser extensometer.

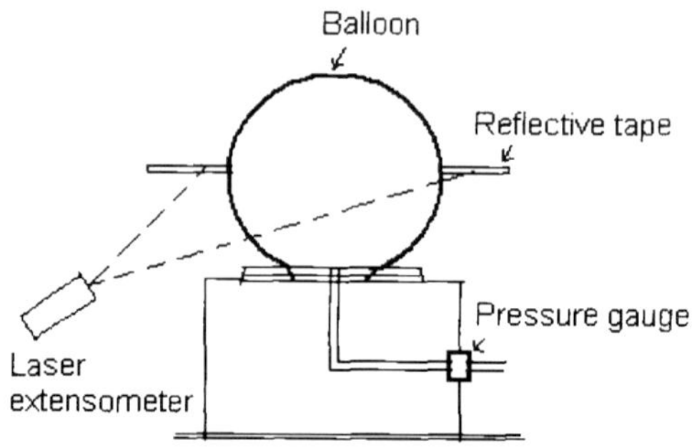

FIG. 5—*Balloon type equibiaxial extension equipment.*

Tear Test

The fracture energy or tearing energy was determined according to ASTM D 624-91: Standard test method for tear strength of conventional vulcanized rubber and thermoplastic rubbers. The trouser test piece (designated as Die T in the ASTM standard, see Fig. 6) was chosen in this study because it is particularly convenient for calculating the tearing energy in order to give more fundamental results.

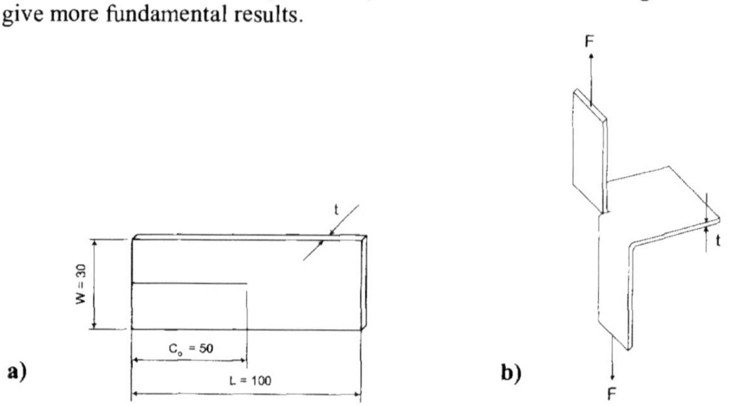

FIG. 6—*Trouser test specimen: (a) undeformed state; (b) extended state.*

Results and Discussion

Figure 7 shows the typical curves of force versus vertical displacement recorded during the puncture tests. These curves are not linear and show a maximum force F, and a maximum vertical displacement H at puncture. Figure 8 shows the effect of the angle of the probe tip on the puncture resistance of neoprene sheet of thickness t = 0.78 mm. The conical probes had the same rod diameter d_2 = 1.05 mm and the same tip diameter d_1 = 0.05 mm, but different tip-angles (see Fig. 4*d*). The tip angles of the probes were α = 13°, 30°, 45°, and 60° (α = 13° is the probe used in ASTM F 1342).

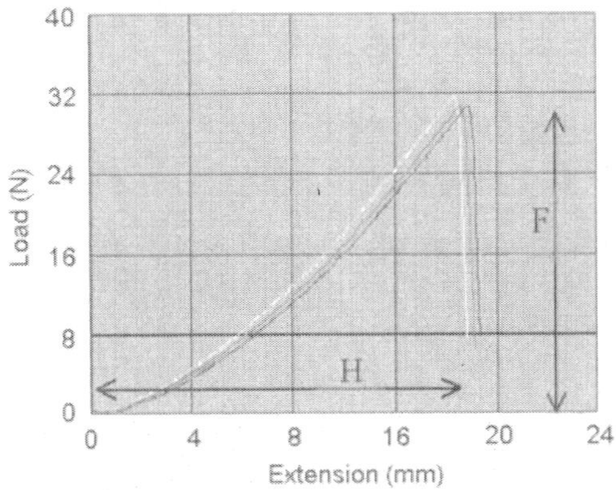

FIG. 7—*Typical force - vertical displacement relationships for nitrile (t = 0.30 mm).*

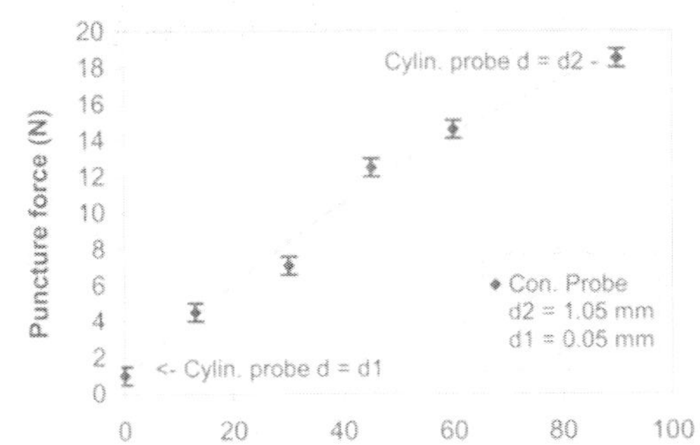

FIG. 8—*The dependence of puncture force on the probe tip angle of the conical probe (Neoprene 0.78 mm thick).*

The results show that the puncture force increases almost linearly with the probe tip angle. In the same figure, the results obtained with two cylindrical probes of diameter 0.05 mm and 1.05 mm are also shown. These cylindrical probes correspond to the cases of α = 0° and α = 90°, respectively. The measurements show that, within the experimental error, the maximum puncture force obtained with the cylindrical probes falls on the same curve as those obtained with the conical probes. These results suggest that the puncture resistance of elastomeric membranes to a conical probe can be determined from the resistance to a cylindrical probe. Therefore, the puncture resistance to the conical probe used in ASTM F 1342 can also be determined from the resistance to the corresponding two cylindrical probes. These results also suggest that puncture by the conical probe and by cylindrical probe are controlled by the same material parameters. In order to determine the intrinsic material parameters in puncture, cylindrical probes with flat tips of different diameters were used.

Figure 9 shows the normalized maximum force *F/t* as a function of the probe diameter for neoprene sheets of three different thicknesses. It can be seen that the maximum puncture force is proportional to the sheet thickness and the probe diameter. Measurements on other elastomers, shown in Fig. 10, confirm the same correlation. Also, puncture tests were performed on nitrile 0.30 mm thick, with two different sample holder hole diameters D_h (see Fig. 3), and the results are presented in Fig. 11. These results demonstrate that the effect of sample holder size on the puncture force is negligible. These results suggest that puncture resistance is controlled by a single material parameter (mechanical property), independent of sample thickness, the probe tip diameter, and the sample holder diameter $\left(\frac{F}{td} = const.\right)$. An attempt is made to calculate one important material parameter: the failure strain when puncture occurs.

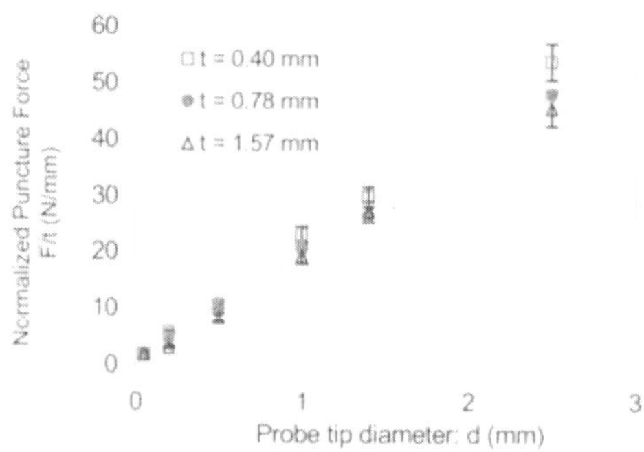

FIG. 9—*Normalized puncture force as a function of probe tip diameter for different thicknesses of neoprene (cylindrical probes with flat tips).*

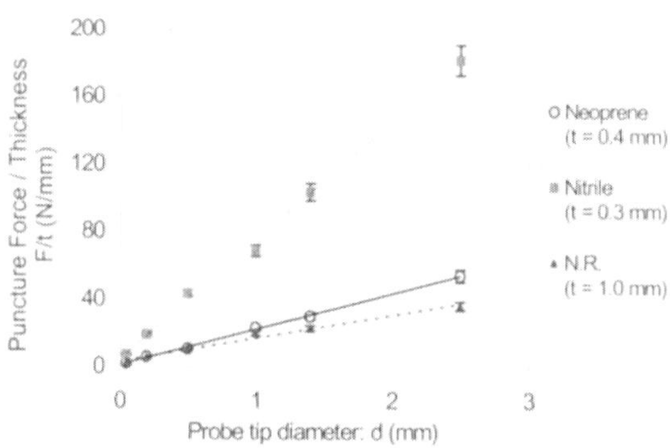

FIG. 10—*Relationship between normalized puncture force (F/t) and probe tip diameter.*

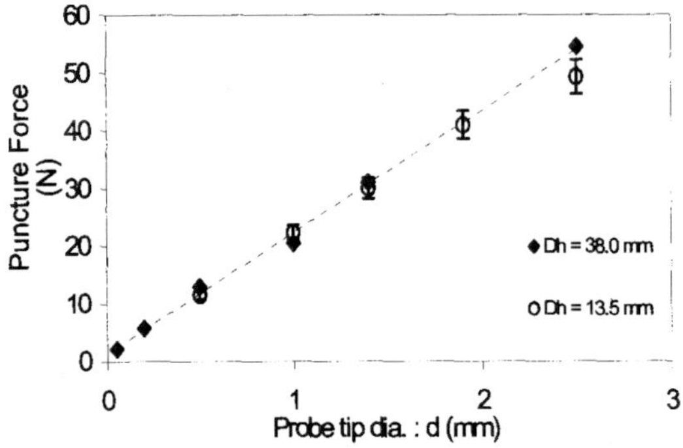

FIG. 11—*Puncture force-probe tip diameter relationship for different sample holder hole diameters D_h (material: nitrile 0.30 mm thick).*

Observations of the deformations of samples before puncture revealed that at the probe tip, the rubber membrane adheres to the cylindrical geometry of the probe over a specific distance from the probe tip as shown in Fig. 12. The results suggest that puncture is caused by local deformations of the membrane at the probe tip and explain why the puncture force is not dependent on the sample holder size but only on probe diameter. After puncture, a hole in the membrane with a specific diameter is observed, and a small piece of the rubber material is cut out as shown in Fig. 13. Figure 14 shows the micrographs of the drop-out disks of neoprene and nitrile after puncture. The diameter of these disks d_f depends on the probe diameter d and is always smaller than that of the probe ($d_f < d$). The diameters of the drop-out disks were measured with an optical microscope, and the results are presented in Table 1. The results show that, regardless of the probe diameter, the ratio d/d_f is constant for each material.

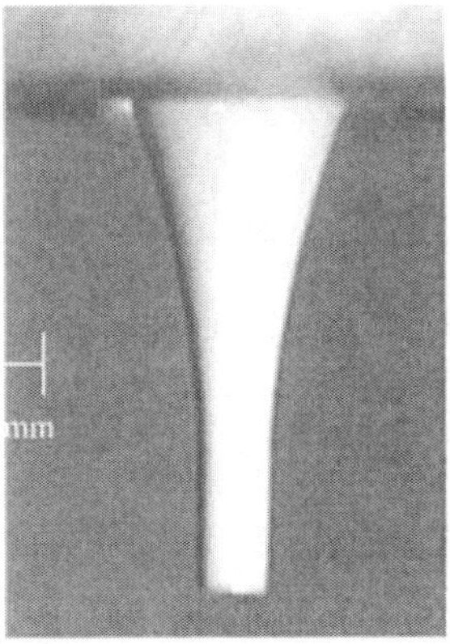

a) Nitrile

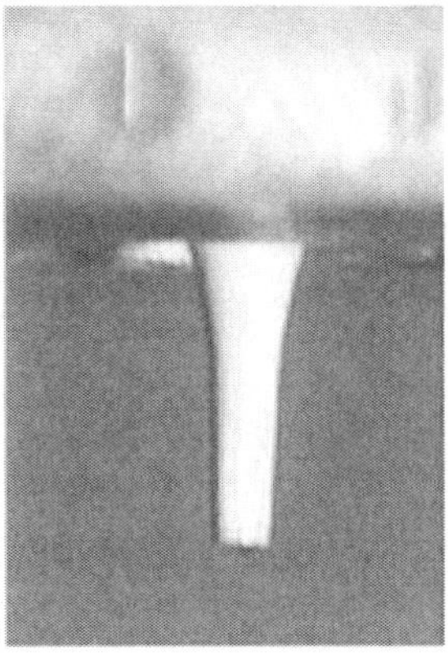

b) Natural rubber

FIG. 12—*Deformation of samples before puncture.*

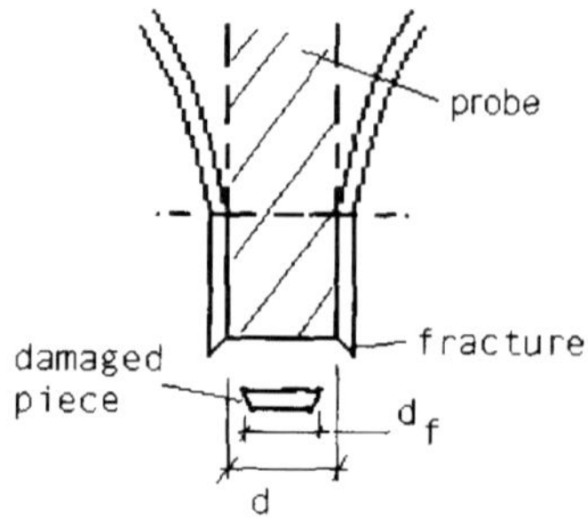

FIG. 13—*Sketch of fracture.*

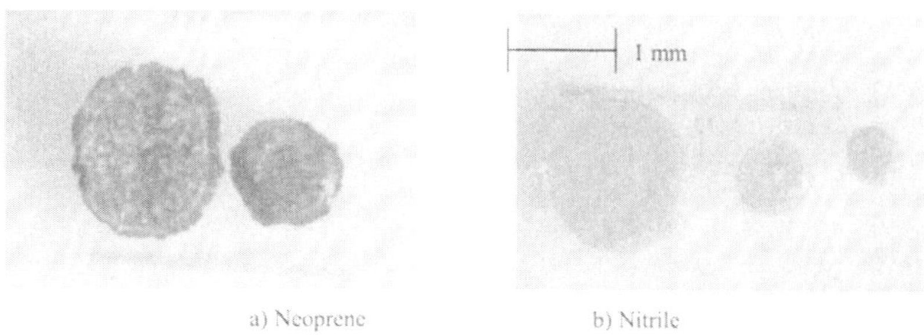

a) Neoprene   b) Nitrile

FIG. 14—*Damaged drop-out pieces of samples (by microscope 20×).*

TABLE 1—*Relationships between probe tip diameter and drop-out disk diameter.*

Material	Nitrile			Neoprene		N.R.
d (mm)	1.0	1.4	2.5	1.4	2.5	2.5
d_f (mm)	0.43	0.60	1.10	0.55	1.05	0.60
d/d_f	2.32	2.33	2.30	2.54	2.40	4.16
e_r	1.32	1.33	1.30	1.54	1.40	3.16
ε_r	0.84	0.85	0.83	0.93	0.88	1.43

From the deformation observed in Fig. 12, it is reasonable to consider that puncture would take place around the circumferential edge of the cylindrical probe. The diameter of the cut-out disk at the onset of puncture (at maximum equibiaxial deformation) would correspond to the diameter of the probe. Since the deformation is axisymmetrical, the radial strain ε_r^1 and the tangential strain ε_t^1 of the cut-out disk can be calculated. The extension ratio λ_r in the radial direction is:

$$\lambda_r = d/d_f \tag{1}$$

and the extension ratio λ_t in the tangential direction is:

$$\lambda_t = \frac{\text{Circumference of cut - out disk at puncture (at maximum deformation)}}{\text{Circumference of cut - out disk (under undeformed state)}} = \frac{\pi d}{\pi d_f} = \frac{d}{d_f}$$

The corresponding engineering strains and true strains are therefore:

$$e_r = e_t = \lambda_t - 1 = d/d_f - 1 \qquad \text{(engineering strains)} \qquad (2)$$
$$\varepsilon_r^{\,l} = \varepsilon_t^{\,l} = \mathrm{Ln}\,(\lambda_t) = \mathrm{Ln}\,(d/d_f) \qquad \text{(true strains)}$$

To verify the above assumption, the diameters of the cut-out disks were measured with an optical microscope, and the results are presented in Table 1.

Regardless of the probe diameter, the ratio d/d_f is constant for each material and is very close to the maximum strains of the elastomer membrane, measured from the top surface of the probe.

In order to verify whether puncture is caused by stress concentration around the circumferential edge of the cylindrical probe with a flat tip, puncture tests were performed using a cylindrical probe with a semi-hemispherical tip. Figure 15 shows that the deformation at the top surface remains equibiaxial (Fig. 15*b*), and the deformation pattern at the probe tip is similar to that of the flat-tip probe.

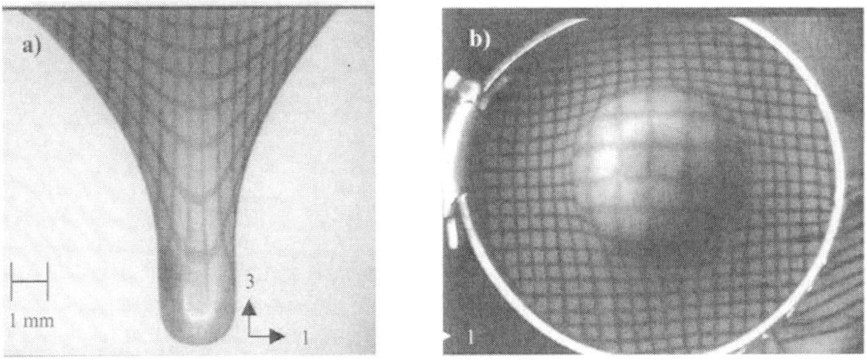

FIG. 15—*Deformation of sample around rounded-tip probe* (a)*; deformation at rounded tip* (b) *($e_1 = e_2$, material: nitrile, probe tip: d = 2.5 mm).*

Table 2 shows the comparisons of the maximum force and deformation at the probe tip for flat and hemispherical tips. It is surprising to observe that the hemispherical probe tip gives the same maximum puncture force as the flat tip. Measurements of the maximum strain (Table 2) also show that there is no difference between these two probe-tip geometries. The result suggests that the puncture failure strain in the elastomer membrane is not due to stress concentration around the edge of the flat tip. Within the range of probe diameters investigated in this work (from 0.1–2.5 mm), puncture is controlled by a maximum local deformation (maximum strain) that is independent of the probe diameter.

To find the relationship between tensile failure strain and puncture failure strain, tensile tests were carried out, and the results are presented in Tables 3 and 4. The puncture failure stress in Table 4 was calculated from the puncture force. Since the contribution of the friction force is minor (verified by puncture tests with lubricant) and because the rubber membrane adheres to the puncture probe, the total force in the membrane could be considered to be equal to the puncture force. So that $F = \sigma \cdot t \cdot \pi d$, and the puncture failure stress values presented in Table 4 were calculated from: $\sigma = F/(t \cdot \pi d)$.

TABLE 2—*Test results (and their standard deviation, SD) for two types of puncture probes (d = 1.0 mm).*

Material	Neoprene		Nitrile		N.R.	
Probe tip type	Flat	Rounded	Flat	Rounded	Flat	Rounded
Puncture force F_p (N)	13.5 ± 0.4	13.6 ± 0.5	21.2 ± 1.4	22.0 ± 0.9	21.0 ± 1.1	20.6 ± 1.6
Max. strain	1.46 ± .13	1.40 ± .17	1.32 ± .08	1.35 ± .10	3.18 ± .21	3.05 ± .23

TABLE 3—*Failure engineering strain and true strain of tensile and biaxial balloon tests compared to puncture tests (in parentheses: SD).*

Material		Nitrile	Neoprene	N.R.
Failure engineering strain (%)	Tensile	410 (22)	376 (18)	971 (51)
	Biaxial	239 (17)	186 (16)	493 (42)
	Puncture	132 (11)	146 (7)	318 (23)
Failure true strain (%)	Tensile	163 (10)	156 (8)	237 (14)
	Biaxial	122 (9)	105 (9)	178 (15)
	Puncture	84 (8)	90 (5)	143 (12)

TABLE 4—*Failure engineering stress and true stress of tensile and biaxial balloon tests compared to puncture tests (in parentheses: SD).*

Material		Nitrile	Neoprene	N.R.
Failure engineering stress (MPa)	Tensile	41(4)	15 (2)	30 (3)
	Biaxial	80 (7)	26 (3)	54 (6)
	Puncture	55 (7)	18 (2)	25 (3)
Failure true stress (MPa)	Tensile	210 (22)	71 (7)	325 (31)
	Biaxial	270 (29)	73 (9)	317 (35)
	Puncture	128 (18)	45 (7)	104 (11)

The puncture strength appears to be much smaller than that measured by tensile tests. Since puncture involves a biaxial state of stress, it is relevant to verify whether these discrepancies can be due to the effect of biaxial stress. To determine the maximum equibiaxial strength, the elastomer membranes were inflated by air until they burst. The pressure was measured and controlled by a pressure gauge. The maximum diameter of the balloon before bursting was recorded by an MTS LX 500 laser extensometer. The true strain and stress failure were calculated by [11, 12]:

$$\sigma = \lambda^2 RP/2t \tag{3}$$

with $\lambda^2 = S_{ball}/S_0$

where R is the maximum radius of the balloon at bursting, P is the maximum pressure, t is the initial thickness, S_{ball} is the maximum surface area of the balloon at bursting, and S_0 is the initial surface of the sample.

The results show that puncture strength is also smaller than the biaxial strength and refutes the hypothesis that puncture is controlled by an equibiaxial strength. The results suggest that puncture is controlled by local deformations of the membrane at the probe tip and explain why puncture stress is much smaller than the tensile and equibiaxial strengths.

Material Puncture Resistance to Medical Needles

As mentioned previously, puncture caused by actual sharp and pointed objects is different from that by the conical probe or cylindrical probe in the existing test methods. For example, with sharp medical needles, the puncture resistance is much lower than that measured by ASTM F 1342 [1]. In this section, the puncture resistance of a neoprene material was characterized using medical needles as puncture probes.

Puncture by conical probe happens suddenly when the strain at the probe tip reaches the failure value and no fracture surface was created before puncture occurred. However, the puncture process by medical needle happens gradually; after needle tip enters the sample thickness and creates fracture surfaces, it continues to penetrate deeper, widens the fracture surfaces, which makes the force continue to increase. A piece of aluminum foil placed underneath the sample was used to complete an electrical circuit between the foil and the needle in order to detect the emergence of the needle tip when puncture occurred (a hole was created through the sample thickness). See Fig. 16 for a sketch of the puncture test.

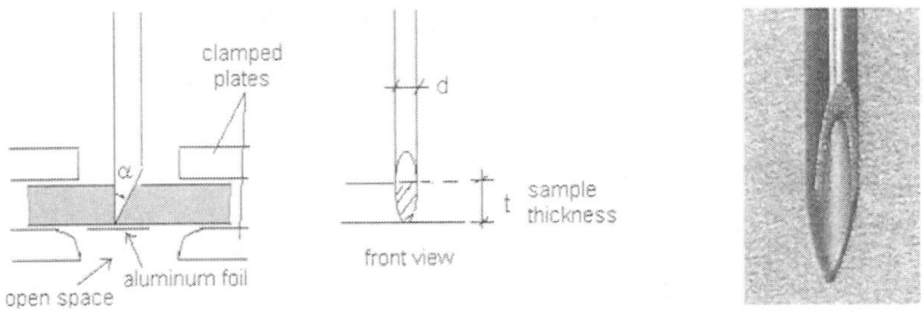

FIG. 16—*Sketch of puncture test by medical needle.*

The fracture energy G_c is one important mechanical property of the material (besides modulus, tensile strength). An attempt will be made to determine the relationship between the material fracture energy G_c and the energy of the new fracture surfaces G_s created by puncture by medical needle. Using the ASTM D 624 tear test, the fracture energy G_c of the neoprene material used in this study was characterized. The value obtained for G_c is 8.1 (mJ/mm^2), with tearing speed equal to puncture speed in this study.

As mentioned in fracture mechanics, the energy released during cracking supplies the energy to create the new fracture surfaces, G_s, which can be calculated using the following equation:

$$W = G_s A + W_r \tag{4}$$

where W is the total strain energy immediately before puncture, W_r is the remaining strain energy after puncture, computed from the force-displacement curve (Fig. 17) and the area A of one created fracture surface, which is measured with an optical microscope (Fig. 18).

The calculated values G_s obtained with three needles of 0.35, 0.65, and 0.80 mm diameter for neoprene with two thicknesses (0.40 and 1.57 mm) are presented in Table 5. The calculated values G_s, 8.8–10.4 mJ/mm^2, from the puncture test in Table 5, are close to the fracture energy from the tear test mentioned above, G_c = 8.1 (mJ/mm^2). This result suggests that the puncture resistance to medical needles relates to the fracture energy of the material, whereas the puncture resistance to cylindrical probes relates to the material's tensile strength and failure strain.

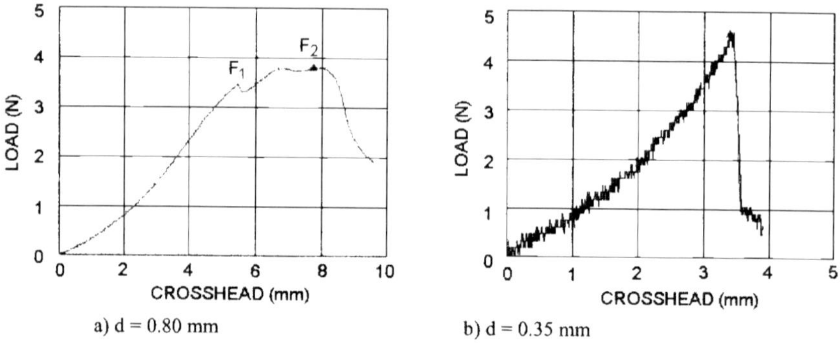

FIG. 17—*Force-vertical displacement relationships for medical needles on neoprene 0.78 mm thick.*

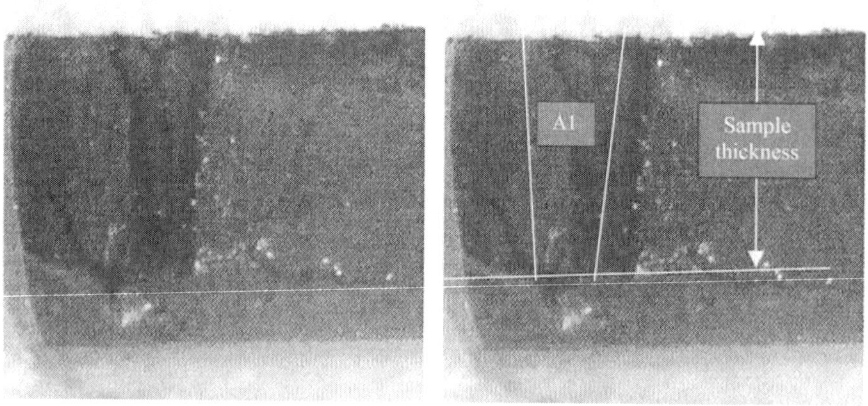

FIG. 18—*Fracture surface created by needle (by optical microscope 20×).*

TABLE 5—*Calculated energy of the fracture surfaces created by puncture by medical needle.*

Probe diameter: d (mm)	0.35	0.50	0.65
Sample thickness (0.40 mm)	8.8 (1.6)	9.1 (2.2)	10.4 (2.4)
Sample thickness (1.57 mm)	9.3 (1.9)	9.2 (2.0)	10.2 (1.8)

Unit: mJ/mm^2. In parentheses: SD. Each value is average of four tests.

The results presented suggest that puncture resistance to medical needles involves cutting. Some studies on cutting [13–15] have related the cutting force (or cutting energy) and the fracture energy of the material. Consequently, puncture with a sharp medical needle could also relate to the fracture energy as demonstrated above.

To confirm the different mechanisms between puncture by sharp medical needle and by conical/cylindrical probe, the tests were carried out on aged neoprene. The results in Fig. 19 show that there are different effects on puncture resistance for two probe types. For the conical probe, the puncture resistance is decreased with aging time, whereas the puncture resistance of the medical needle is increased. This is because for rubber, thermo-oxidative degradation is prominent, the neoprene gradually becomes harder, and the capacity of elastic deformation decreases. This leads to a decrease in the failure strain along with an increase in modulus [16]. Results from the previous section show that the puncture resistance of conical/cylindrical probe is related to material tensile strength and failure strain. Thus, as the failure strain is decreased with aging time, the puncture resistance of the conical/cylindrical probe is also decreased.

As shown previously, because the medical needle is very sharp, the puncture is likely a cut-through the thickness of the sample. In another study [18], the cutting resistance of rubber is found to be proportional to the modulus or hardness. When aging, neoprene becomes harder (increase in modulus), which leads to an increased cut resistance as well as puncture resistance to the medical needle.

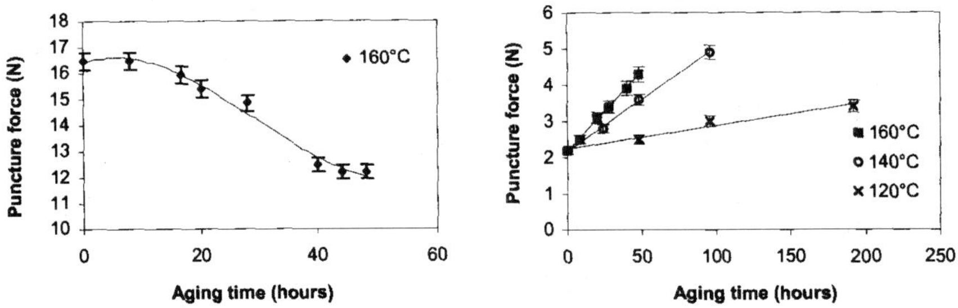

a) Conical probe in ASTM F 1342

b) Medical needles d = 0.65 mm

FIG. 19—*Variation of puncture force of neoprene 0.78 mm with aging time at various aging temperatures.*

Conclusions

It has been shown in this work that the geometry of the conical probe used in ASTM Standard F 1342 strongly affects the puncture characterization results of elastomer membranes. The results demonstrate that simpler cylindrical probes provide a more straightforward characterization of puncture and also eliminate the complex effect of probe-tip geometry. The puncture strength is much smaller than both the tensile strength and the biaxial strength of the elastomer membrane. After puncture, a hole in the membrane with a specific diameter is always observed, with a small disk cut out from the rubber membrane. The diameter of the cut-out disks depends on the probe diameter and is always smaller than that of the probe. Regardless of the probe diameter, the "hole diameter/probe diameter" ratio is constant for each material and is very close to the maximum strains in puncture of the elastomer membrane, measured from the top surface of the probe. The puncture failure strain in the elastomer membrane is not due to stress concentration around the edge of the flat tip and is the same for rounded tips. Puncture of elastomer membranes is controlled by a maximum local deformation that is independent of the probe diameter.

For sharp medical needles, the puncture process happens gradually and involves cutting, which relates to the fracture energy of the material, whereas the puncture resistance to cylindrical probes relates to the material's tensile strength and failure strain.

References

[1] Leslie, L. F., Woods, J. A., Thacker, J. G., Morgan, R. F., McGregor, W., and Edlich, R. F., "Needle Puncture Resistance of Medical Gloves, Finger Guards, and Glove Liners," *Journal of Biomedical Materials Research*, Vol. 33, No. 1, 1996, pp. 41–46.

[2] Salkin, J. A., Stuchin, S. A., Kummer, F. J., and Reininger, R., "The Effectiveness of Cut-Proof Glove Liners: Cut and Puncture Resistance, Dexterity, and Sensibility," *Orthopedics: Thorofare*, Vol. 18, No. 11, 1995, pp. 1067–1071.

[3] Hewett, D. J., *Protocol for the Puncture Resistance of Medical Glove Liners*, Protective Equipment Section, National Institute for Occupational Safety and Health, USA, 1993, 37 pp.

[4] Stevenson, A. And Malek, K. A. B., "On the Puncture Mechanics of Rubber," *Rubber Chemistry and Technology*, Vol. 67, No. 5, 1994, pp. 743–760.

[5] Murphy, V. P. and Koerner, R. M., "CBR Strength (Puncture) of Geosynthetics," *Geotechnical Testing Journal*, Vol. 3, ASTM International, West Conshohocken, PA, 1988, pp. 167–172.

[6] Narejo, D., Koerner, R. M., and Wilson-Fahmy, R. F., "Puncture Protection of Geomembranes, Part II: Experimental," *Geosynthetics International*, Vol. 3, 1996, pp. 629–653.

[7] Wilson-Fahmy, R. F., Narejo, D., and Koerner, R. M., "Puncture Protection of Geomembranes, Part I: Theory," *Geosynthetics International*, Vol. 3, 1996, pp. 605–627.

[8] Ghosh, T. K., "Puncture Resistance of Pre-Strained Geotextiles and its Relation to Uniaxial Tensile Strain at Failure," *Geotextiles and Geomembranes*, Vol. 16, 1998, pp. 293–302.

[9] ASTM Standard F 1342: "Standard Test Method for Protective Clothing Material Resistance to Puncture," *Annual Book of ASTM Standards*, Vol. 11.03, ASTM International, West Conshohocken, PA, 1999, pp. 1102–1104.

[10] ASTM Standard D 412: "Standard Test Methods for Vulcanized Rubber and Thermoplastic Elastomers – Tension," *Annual Book of ASTM Standards*, Vol. 09.01, ASTM International, West Conshohocken, PA, 1998.

[11] Yoshihiro, Y., Sueo, K., Yuji, T., and Wataru, S., "Biaxial Extension Strength of Carbon Black Reinforced Rubber," *Nihon Gomu Kyokaishi*, Vol. 67, No. 8, 1994, p. 576.

[12] Kawataba, S., "Fracture and Mechanical Behavior of Rubber-Like Polymers under Finite Deformation in Biaxial Stress Field," *J. Macromol. Sci.- Phys.*, B8(3–4), 1973, pp. 605–630.

[13] Wang, C. and Gent, A. N., "Energy Dissipation in Cutting and Tearing Rubber," IUPAC Polymer Symposium: Functional and High Performance Polymers, Taiwan, 1994, pp. 629–631.

[14] Cho, K. and Lee, D., "Viscoelastic Effects in Cutting of Elastomers by a Sharp Object," *Journal of Polymer Science Part B: Polymer Physics*, Vol. 36, No. 8, 1998, pp. 1283–1291.

[15] Gent, A. N., Lai, S-M, Nah, C., and Wang, C., "Viscoelastic Effects in Cutting and Tearing Rubber," *Rubber Chemistry and Technology*, Vol. 67, No. 4, 1994, pp. 610–619.

[16] Budrugeac, P., "Accelerated Thermal Aging of NBR and Other Materials under Air or Oxygen Pressures," *Die Angewandte Makromolekulare Chemie*, Vol. 247, 1997, pp. 19–30.

[17] Mott, P. H. and Roland, C. M., "Aging of natural rubber in air and seawater," *Rubber Chemistry and Technology*, Vol. 74, 2000, pp. 79–88.

[18] Vu, T. B., "Mechanics and Mechanism of Cutting of Protective Materials," NGA Thesis, Université de Sherbrooke, Quebec, Canada, 2002.

SECTION 5: THERMAL PROTECTIVE ASSESSMENT

Journal of ASTM International, May 2005, Vol. 2, No. 5
Paper ID JAI12114
Available online at www.astm.org

Ned R. Keltner, Ph.D.[1]

Evaluating Thermal Protective Performance Testing

ABSTRACT: Even as better materials are developed for protective clothing, NFPA data indicate the number of burn related fire fighter deaths and severe injuries is increasing. These injuries are unexpected for the most part. As a result, questions arise about Thermal Protective Performance (TPP) ratings and whether performance changes are due to use or the effects of aging. As part of a NIST-sponsored Small Business Innovative Research project, current TPP test techniques were evaluated. Some changes and extensions are suggested.

In TPP tests, dry fabric samples or ensembles are exposed to heat flux of 83 kW/m^2 with nominally 50 % radiative and 50 % convective heat transfer. Heat transmission through the test sample is measured with a copper calorimeter. The **TPP Rating** is the time in seconds required for a 2nd degree burn.
Current problems include:

1) Current test methods overestimate time to 2nd degree burn.
2) No information is provided on maximum potential burn damage.
3) No information is provided on heat transfer or fabric properties.
4) No information is provided on how performance changes with use.

Suggested changes and extensions include:

1) Change the heat source - use a modified radiant protective performance technique.
2) Replace the copper calorimeter with a thermal skin simulant sensor.
3) Estimate burn damage using a two step analysis method.
4) Obtain both dry and damp thermal properties for modeling clothing performance.

KEYWORDS: thermal protective performance, thermal skin simulant, test, burn damage, heat flux

Introduction

The number of burn-related fire fighter deaths and severe injuries is increasing. For the most part, these injuries are unexpected. As a result, questions have risen about what causes the burns, the TPP ratings, and how protective performance changes with use or aging [10].

The current TPP test method used in the United States is part of NFPA 1971 Protective Ensemble for Structural Fire Fighting. In this test method, a combination of quartz infrared lamps and gas burners is used to expose dry material samples to a heat flux of 83 kW/m^2 with nominally 50 % radiative and 50 % convective heat transfer. The Radiant Protective Performance (RPP) test, described in NFPA 1977, for wildland fire fighting clothing, is based upon exposing the clothing materials and the sensor to a 21 kW/m^2 heat flux from just a quartz lamp array.

In both the TPP and RPP tests, heat transmission through the test sample is measured with a copper calorimeter. To evaluate protective clothing performance, data are analyzed using the Stoll Curve to determine the time to 2nd degree burn [17]. This thermal endpoint curve consists of time-temperature points developed for the copper calorimeter using test data on bare skin

Manuscript received 18 November 2003; accepted for publication 29 September 2004; published May 2005.
Presented at ASTM Symposium on Performance of Protective Clothing: Global Needs and Emerging Markets: 8th Symposium on 13 January 2004 in Tampa, FL.
[1] F.I.R.E.S, Inc., Consultant to Ktech Corporation, 1300 Eubank Blvd. SE, Albuquerque, NM 87123-3336.

exposed to uniform heat flux levels. In both tests, a 2nd degree burn is said to occur when the time-varying calorimeter temperature increases above the Stoll Curve.

TPP (or RPP) Rating = the time required for 2nd degree burn under fixed thermal exposure

Various researchers have identified potential problems with the current test methods:

A) The thermal endpoint curve was developed using test data on bare skin exposed to uniform, radiant heat flux levels – heat transfer through a clothing sample is time varying and can involve conductive-convective-radiative modes of heat transfer.
B) The response of the calorimeter and its properties is very different from human skin – the copper calorimeter demonstrates a nearly linear response to uniform heating, while the response of human tissue varies as the $\sqrt{time}$.
C) These methods: 1) do not determine the heat transfer through the clothing, 2) do not factor into the potential burn damage estimates the effect of continued heat transfer from the clothing layer during cooling, and 3) are not useful for evaluating the effects of aging.

As part of a Small Business Innovative Research project sponsored by the National Institute of Standards and Technology – Building Fire Research Laboratory, Ktech Corporation teamed with the Center for Firesafety Studies at Worcester Polytechnic Institute and Beck Engineering Consultants (Okemos, MI) to evaluate current TPP/RPP test techniques and to suggest changes.

Evaluation of Thermal and Radiant Protective Performance Tests

The primary objectives in the design of fire fighter clothing are preventing thermal damage to the skin and minimizing heat stress [10]. To evaluate test methods for protective performance properly, one must have a sound understanding of the effects of heat flux exposure on the skin. Assessments of skin burn damage throughout this work use the onset of a superficial 2nd degree burn. A superficial 2nd degree burn, referred to as a 2nd degree burn, is indicative of complete necrosis of the epidermis and is equivalent to the time to blistering of the skin [16]. This level of damage is used in the current standards NFPA 1971 and 1977. It provides a relatively severe, yet survivable, performance criterion level. The skin will not naturally regenerate if damaged beyond this point.

The normal human skin temperature at the surface is 32.5°C. Thermal damage begins when the temperature is increased above 44°C at the base of the epidermis (basal layer), approximately 80 μm below the surface [17].

Regardless of the mode of application of the heat flux, the temperature rise and therefore the time to a 2nd degree burn are related to the amount of heat absorbed by the skin. Damage to the skin can be represented as a chemical rate process using the Henriques Burn Integral [6]:

$$\Omega = \int_0^t P e^{\left(-\Delta E / R\left(T(t)+273\right)\right)} dt \tag{1}$$

where P is the Frequency Factor (1/s), ΔE is the Activation Energy (J/kmol), and R is the Universal Gas Constant. The value of the integral is based upon the amount of time, t (s), where the temperature, T (°C), of the basal layer is greater than 44°C. The value of the burn damage integral defines a dimensionless injury parameter or thermal damage factor, Ω.

The input variables used in the burn integral to calculate Ω are based upon tests conducted to determine the onset of pain due to thermal exposure and the time to 2nd degree burn; several different sets of values are given in SFPE Engineering Guide [16]. The severity of a burn is determined from the values obtained from evaluating the integral over a given temperature-time history. A value for Ω greater than or equal to 1.0 is equivalent to a 2nd degree burn.

As the temperature increases above 44°C, the rate of damage to the skin increases logarithmically. Depending on the data used in Eq 1, the estimated rate of damage at 50°C is 40 times (Henriques) to 120 times (Weaver&Stoll) greater than the rate of damage at 45°C [16]. Extrapolation of the rate of damage versus skin temperature curve indicates that a 2nd degree burn will occur virtually instantaneously when the basal temperature is 72°C.

Given concerns about the accuracy of TPP ratings and estimates of potential burn damage, a detailed evaluation was made of the TPP and RPP test methods. The results in Refs. [4,5,7,8] have shown:

A) Current test methods overestimate the time to a 2nd degree burn and thus the TPP rating. Tests at WPI indicated the over-estimate of the time ranges from 23–34 %. This results from several factors: 1) Poor heat source fidelity in simulating heat transfer from a structure fire with a quartz lamp/Meeker burner heat source. As a result, the absorbed radiative heat flux depends on the outer shell fabric's radiative properties, which introduces uncertainty into the TPP rating; 2) The Copper Calorimeter distorts heat transfer through the garment because its temperature history is very different than the skin surface response; and 3) The poor performance of the generic flat black paint under the intense thermal loading (e.g., blistering). All of these factors introduce significant uncertainty into TPP ratings.

B) Because analysis stops with determination of the 2nd degree burn time, no information is provided on maximum potential burn damage.

C) Given the limited data reported and that small samples are used in the destructive TPP or RPP tests, the methods cannot be used for other purposes that could benefit fire fighter safety. No information is provided on heat transfer or fabric properties for evaluating protection during other exposures, such as long term exposure at heat fluxes below 83 kW/m^2, calculating heat loss characteristics as part of evaluating thermal stress, or on how protective performance changes during use (e.g., due to absorbing moisture) or with aging (e.g., due to wear and laundering).

Heat Source Fidelity

To produce a flux of 21 kW/m^2 with a quartz lamp array for an RPP test, the temperature of the quartz lamp filaments is approximately 1625 K. This temperature produces radiant heat mainly in the near to middle infrared region, with a spectral peak at approximately 1.8 μm. For the quartz lamp array to produce half of the 83 kW/m^2 total heat flux for a TPP test, the temperature of the quartz lamp filaments increases to approximately 1950 K; this higher temperature shifts the spectral distribution of the radiant heat further toward the near infrared region, with the spectral peak dropping to approximately 1.5 μm. The quartz envelope absorbs most of the radiant energy produced by the heated filament at wavelengths above approximately 4 μm.

To produce the RPP heat flux of 21 kW/m^2 with the resistance heater from a Cone Calorimeter [1], the heater operates at a temperature of approximately 870 K, due to its larger radiating surface. The spectral distribution of the radiant energy covers the near to far infrared region, with 90 % of the energy lying between 1 and 10.5 μm with a spectral peak at approximately 3.3 μm. To produce the radiant half of the TPP heat flux, the temperature increases to approximately 1030 K, which gives a spectral peak at 2.8 μm. Overall, the spectral distribution of the radiant heat from this tubular resistance heater more closely represents the radiative heating from a structure fire than that from the quartz lamps.

Data presented by Quintiere [14] shows that fire fighting coat materials exhibit a strong absorption of infrared energy in a region centered around 3 μm, from about 2.8–3.2 μm.

Figure 1 provides a sample curve for the reflectance of a fire fighting coat shell material, similar to those presented in Quintiere's paper, along with the calculated emissive power per unit wavelength distribution curves for the Quartz Lamp and the Cone Heater. The sample curve of the fabric reflectivity is provided to present the typical shape of the reflectance versus wavelength curve, and it is not meant to be applicable to any particular material. At wavelengths lower than 3 μm, the reflectivity increases significantly (thus decreasing the absorbed heat flux).

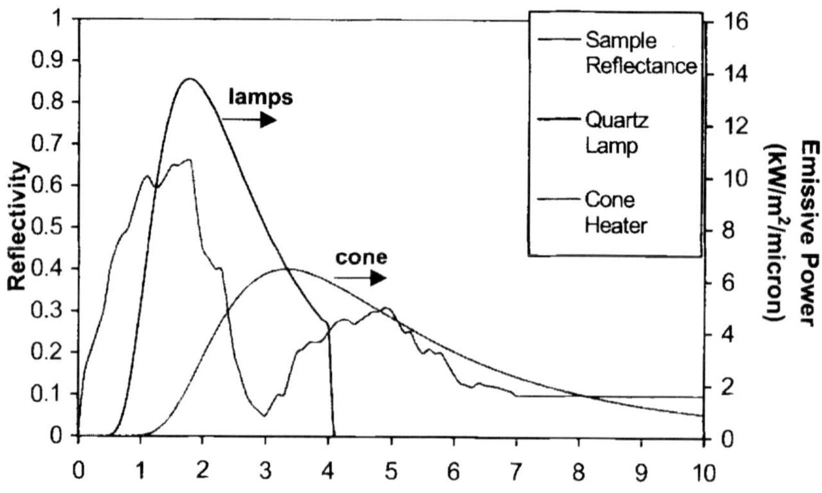

FIG. 1—*Emissive power per unit wavelength of quartz lamp and cone heater in relation to characteristic fire fighting coat material reflectivity.*

Operated at nominally 30 % of rated voltage for the RPP test, the spectral peak of the radiation from the quartz lamp lies at approximately 1.8 μm, which is nearly equivalent to the peak of the characteristic reflectivity curve for this fire fighting coat material. The spectral peak of the Cone Heater at 870 K is nearly equivalent to the minimum of the characteristic reflectivity curve (point of maximum absorptivity) for the sample fire fighting coat material. These differences in the spectral variation of the emissive power of the heat sources when coupled with the sample reflectivity curve of fire fighting coat materials, suggest that the levels of radiant energy absorbed by protective clothing materials will depend on the type of heat source. These differences in absorbed heat flux produce differences (uncertainties) in the prediction of the time to 2nd degree burn between different heat sources and actual fires. The best way to minimize this uncertainty is to use a source that more closely replicates the radiant heat transfer from a fire. One suggestion for this is the heater from the Cone Calorimeter.

For the burner flames in NFPA 1977, part of the heat transfer is convective, and part of it is radiative, coming from water and carbon dioxide bands plus some continuum radiation from soot. As a result, interpreting the effects is more complicated. This issue will need to be addressed as new TPP test methods are developed, and the community needs to understand the level of uncertainty introduced by heat source selection.

Improved Measurement of Potential for Burn Damage

The Copper Calorimeter distorts the transient heat transfer through a garment because its temperature history is very different than the skin surface response. This contributes to current test methods giving longer burn time estimates [4,7].

Thermal skin simulant sensors were evaluated as a possible alternative to address the copper calorimeter problem. Thermal skin simulant sensors are designed to have a thermal inertia parameter ($\sqrt{k\rho\, C_p}$ abbreviated **ThI**) value within 10 % of the value for human tissue. **ThI** controls the surface temperature response during heating typical of protective performance tests. As a result, using these sensors in protective performance tests and thermal property experiments provides a more representative boundary condition on the unheated face of the garment and thus a more accurate measurement of heat transfer through a garment in a protective performance test.

A comparison of the dynamic response characteristics of thermal skin simulant sensors (TSSS) and copper calorimeters (CC) are given below for a step change in heat flux:

$\Delta T(t)=2q\sqrt{t}\,/\sqrt{\pi k\rho C_p}\quad for\,TSSS\quad (2A)$ *versus* $\Delta T(t)=q\,t/\rho C_p L\quad for\,CC\quad (2B)$

ThI = 1500 nominal value for skin [3]; ρC_p = 3 520 000 for copper, which gives

$= q\sqrt{t}/1330$ for TSSS *versus* $= q\,t/5585$ for CC

where q = heat flux (W/m^2), t = time (s), k = thermal conductivity (W / m K), ρ = density (kg/m^3), C_p = specific heat (J/kg K), and L = thickness (m).

Equation 2A is the unit step response (USR) function for a TSSS – the surface temperature rise produced by a step change in the heat flux; Equation 2B is the temperature rise of the CC produced by a step change in the heat flux. For 30-s exposures at the RPP test heat flux level (21 kW/m^2), the calculated temperature rises are:

TPSS: 47 C @ 9 s, 63 C @ 16 s, and 79 C @ 25 s --- surface temperature

CC: 34 C @ 9 s, 60 C @ 16 s, and 94 C @ 25 s --- average plate temperature

The USR is used in two ways. The USR can be used with a time domain, deconvolution technique to develop a complementary unit step response (CUSR) function [15]. The CUSR is a deconvolution type of digital filter which solves an inverse problem for thermal protection tests. Namely, what heat flux history produced the measured temperature history? This digital filtering process converts the measured surface temperature history from either sensor into a heat flux history.

Once a heat flux history is obtained with one sensor, it can be used with the USR of a different sensor and a time domain, convolution technique to predict the temperature history of the second sensor. As discussed later, this approach was used to compare data from a cast resin mannequin sensor (ThermoMan) with the traditional copper calorimeter used in the TPP and RPP tests.

The heat flux history can be used directly with some burn models to estimate potential damage [16]. It can be used with a skin model, e.g., Knox [9], to predict the surface temperature history or the basal layer temperature needed for Henriques' Burn Damage Integral.

Two questions generally come up when using deconvolution procedures – computational time, which generally runs from the start of the experiment, and the effects of noise. Given the sampling rates of < 10 Hz typical of many protective performance tests and the relatively short test durations, these calculations are suitable for on-line processing of data during an experiment. For long exposures at low heat flux levels, a fast and accurate truncation procedure has been developed [8].

The digital filtering algorithm is developed from an analytical solution. It works best with high accuracy on noise free data, i.e., manufactured data. Unfortunately, experimental data is never noise free. The question addressed is whether the deconvolution filtering process amplifies or attenuates the noise? Given that the response model has a low pass system response, it appears the deconvolution will amplify the noise; however, the amplification was not excessive. It was demonstrated that as long as the same smoothing filter was used, the order was not important – one should smooth the measurements and then calculate the heat flux or calculate then smooth the results.

To generate heat flux data, the deconvolution, digital filter for a Macor TSSS was used to analyze data generated at Worcester Polytechnic Institute (WPI). In this experiment, the Macor TSSS was exposed to an incident heat flux of 4 kW/m^2. The temperatures from a surface thermocouple and a thermocouple nominally 2.5 mm below the surface were recorded at 10 Hz.

Figure 2 shows the heat flux calculated directly from the measured temperature and then smoothed with a Gaussian kernel function included in the MATHCAD software.

The calculations in Fig. 2 used a measured **ThI** value of 1600 $W\sqrt{s}/m^2 K$ for Macor; this is lower than the manufacturer's thermal property data indicating a value of approximately 1750. Based on Eq 2A, the surface temperature rise will be approximately 6 % lower than skin. Having accurate property data is important for obtaining good burn damage estimates. Errors in the **ThI** value will introduce errors in the heat flux history calculated by unfolding the measured temperature history.

A contentious issue at the ASTM-sponsored "Seventh Symposium on Performance of Protective Clothing: Issues and Priorities for the 21st Century" was the performance, utility, and accuracy of thermal sensors used in TPP tests. Current work has shown that for purely radiative exposures as in the RPP tests or the radiant calibration in TPP tests, comparable heat flux measurements can be obtained using the different methods. However, the different sensor designs will perform differently in actual thermal protective performance tests where there are time-varying, multiple modes of heat transfer.

For sensors like the Macor TSSS or the cast-resin Thermo-Man sensor, the surface temperature response depends on the **thermal inertia parameter** with in-depth response also dependent on the **thermal diffusivity**; for the Copper Calorimeter, Eq 2B shows the response depends on the **thermal mass**. For a given exposure, the different sensor temperature histories (i.e., unheated face thermal boundary conditions) affect the heat transfer through the garments and thus the protective performance ratings. This is especially true when moisture is present.

Using different methods to interpret this heat flux data, such as the Stoll Curve or the Henriques Integral, also leads to different ratings (estimates) of thermal protective performance.

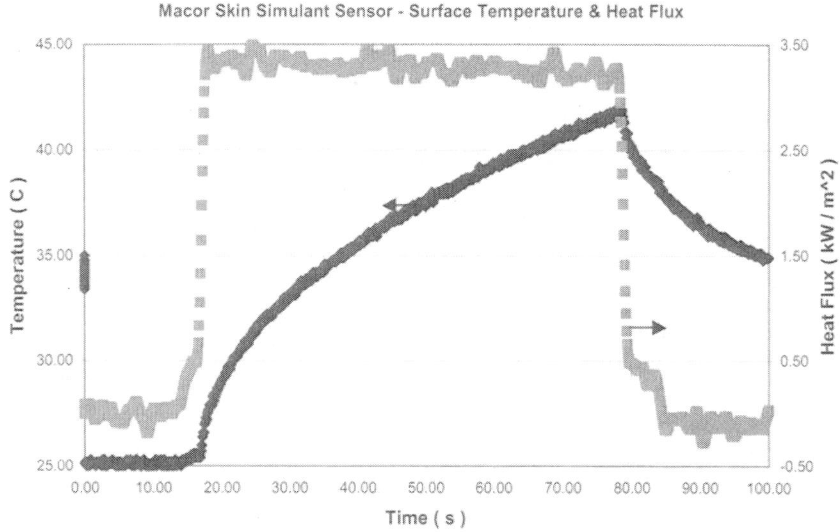

FIG. 2—*Heat flux from surface temperature history using a deconvolution digital filter.*

An analysis of sensor data from mannequin tests provides one view of how sensor design plus the thermal inertia and thermal diffusivity properties affect the results. The 'skin simulant' sensor used in the Dupont Thermo-Man has a copper-constantan thermocouple embedded 5 mils below the surface; the cast resin sensor is manufactured by Composites USA. Dr. Roger Parry (Dupont) provided four Thermo-Man sensors for evaluation, along with some thermal property and flash fire data.

For the Thermo-Man sensors, thermal property data indicated the thermal diffusivity is 30–45 % lower than literature values for human tissue [3]. Combined with a nominal thermocouple location 127 micron (5 mils) below the surface, this suggested the sensor would respond slower than the 80 micron (3 mil nominal) depth of the basal layer in human tissue, which is normally used in estimating burn damage. Estimates for the time delay from initiation heating range from 3 milliseconds for human tissue at the basal layer depth to 10 milliseconds for the thermocouple location in the Thermo-Man sensor.

In addition to the thermal properties, the USR or transient response of this sensor to a step change in heat flux depends on several design features. They include the diameter of the thermocouple wire, the size of the thermocouple bead relative to the wire diameter, the number of wire diameters for which the thermocouple leads are mounted parallel to the heated surface, and the uncertainty in the location of the thermocouple, nominally 127 microns below the front surface. Preliminary tests with the Thermo-Man sensors showed the response was slower than would be expected at the 127 micron depth based on thermal diffusion.

For the Thermo-Man sensors, the thermal inertia parameter is only 35 % that of human tissue. As a result, the front surface temperature rise rate will be up to 3× that of human tissue in cases of high radiant heat fluxes, such as flash fires or arc exposures or TPP tests.

For a step heat flux type of exposure, the very rapid surface temperature rise of the Thermo-Man cast resin sensors overwhelms the slower thermal diffusion process. As a result, the predicted temperature at the 127 micron (nominally 5 mil) thermocouple depth is much higher than that predicted at the 80 micron depth in human tissue. The percentage quickly reaches 200 % and climbs to about 245 % after 40 s of exposure. If there was constant heat transfer through the protective clothing, a relatively simple scaling relation might be useful for estimating burn damage from the sensor data.

However, heat transfer through protective clothing is generally non-uniform. As a result, the situation is much more complex. Heat transfer to sensors used in mannequin tests would depend on whether the sensor is directly exposed to heating by the flames or if it is underneath protective clothing. In general, simple scaling will be nearly impossible because the transient response of the sensor is convolved with the time-varying, location-dependent heat flux.

Following the initial mannequin analyses, temperature measurement histories from a flash fire test were analyzed. A menu driven, inverse heat conduction code, IHCP1D [2], used the in-depth temperature measurements from a 3.5-s exposure to calculate temperature and heat flux histories at the surface as well as the heat flux passing through the plane of the thermocouple. The results are shown in Fig. 3.

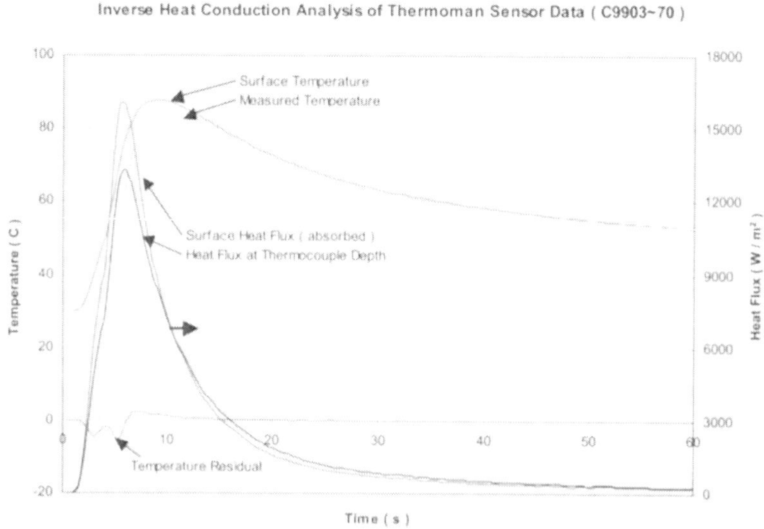

FIG. 3—*Temperature and heat flux calculated with IHCP1D for the surface and thermocouple locations in a Thermo-Man (mannequin) sensor. (Data courtesy of Dr. Roger Parry, Dupont).*

The predicted peak absorbed heat flux was approximately 16 kW/m^2 and occurs at approximately 6 s. The peak predicted surface temperature rise for the cast resin sensor was 55 K; it occurred at 9–9.5 s, and the surface temperature dropped steadily from there.

As expected, the surface temperature responds faster than the in-depth thermocouple location. Due to energy storage in the heated resin, the peak surface heat flux is approximately

20 % higher than the heat flux passing through the thermocouple plane (5 mils below the surface).

Stoll and Chianta used constant heating to develop their burn damage correlation. As can be seen in the highly peaked curves in Fig. 3, heat transfer through protective clothing is not constant. As a result, part of the analysis involved calculating skin surface and copper calorimeter temperatures using the surface heat flux history obtained from the ThermoMan sensor with the inverse heat conduction code (IHCP1D). The results are shown in Fig. 4.

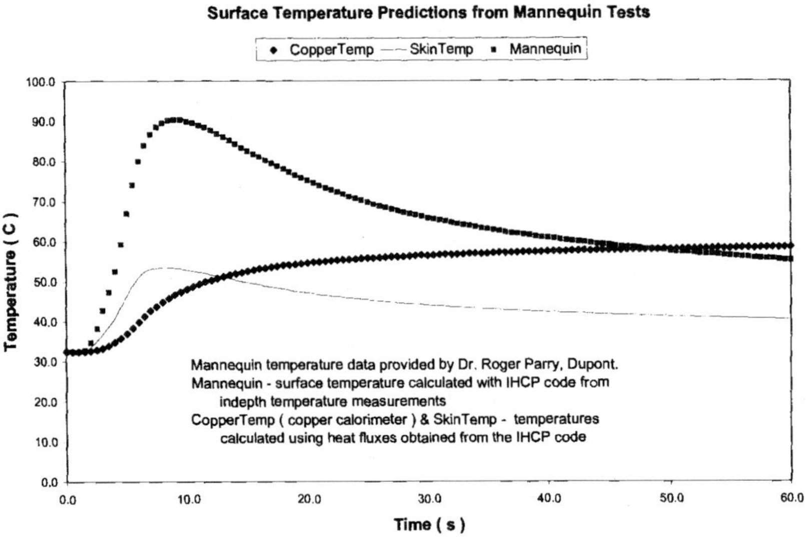

FIG. 4—*Estimated temperatures for a mannequin test (initial temperature = 32.5°C).*

The peaked shape of the calculated skin surface temperature was similar to that of the mannequin sensor; however, the peak temperature rise between 8 and 8.5 s was only 21 K (38 % of 55 K for the mannequin) due to the skin's higher thermal inertia parameter. From the predicted skin temperature curve, burn damage would begin to occur around 5 s. For the copper calorimeter, the shape of the temperature history was different. It continued to rise over a period of 60 s; the temperature rise of the copper calorimeter was approximately 15 K at the time the other temperatures were peaking. Comparing the copper calorimeter data in Fig. 4 with the Stoll and Chianta burn damage tables from ASTM F 1060, the estimated time for a 2nd degree burn is between 8.5 and 9 s.

In the thermal protective performance tests, the copper calorimeter temperature lags at the tissue surface temperature, while the mannequin sensor far exceeds it. As a result, both sensors alter heat transfer through the garment by changing the unexposed face boundary condition.

Due to difficulties finding a material that matches temperature response of human tissue, a two step approach for estimating potential burn damage was developed at WPI [4,5]. One part involves using the skin simulant sensors whose surface temperature more closely tracks the temporal surface temperature response of human tissue. This gets the heat transfer in the experiment as close as possible to the real situation and provides a more accurate heat transfer measurement.

The absorbed heat flux data can be obtained from the analysis of experimental data from any sensor using either a deconvolution filter approach or the inverse heat conduction code. This surface heat flux history can then be used with a numerical technique, either Henriques Burn Damage Integral [6] or a numerical skin model [9,16,20] for estimating potential burn damage during both heating and cool-down.

An example of the technique using the Henriques Burn Damage Integral with data from an RPP test of a single layer of 7.5 oz/sq yd PBI shell material is shown in Fig. 5. It shows that very little of the damage actually occurs before the heat source is removed. The exposure time was limited to 8.7 s because results from a clothing/skin 1-D model indicated this was the minimum exposure that would result in a 2nd degree burn. The Henriques Damage Integral is a rate function equation, such that damage rate increases as the basal temperature increases.

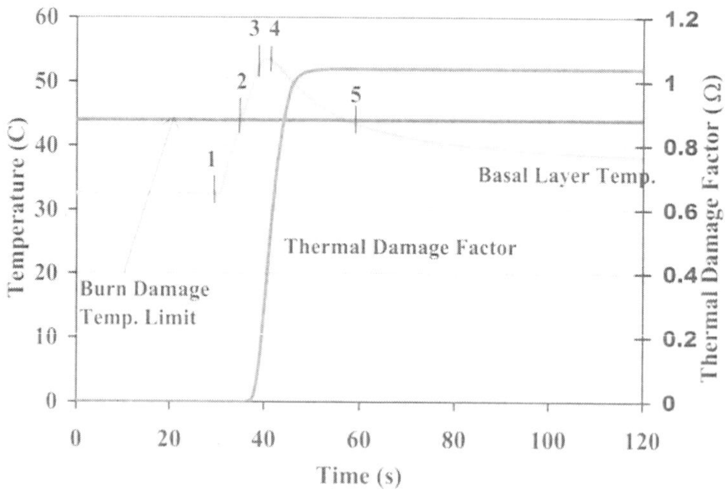

FIG. 5—*Basal layer temperature and Henriques Damage Integral function during evaluation of PBI 7.5 oz/sq yd. Points of interest on curve in order from left to right:*

1. *Exposure begins (29 s).*
2. *Basal layer temperature rises above 44°C (35.1 s).*
3. *Exposure source removed (37.7 s).*
4. *Skin begins to cool (40.3 s).*
5. *End of damage to skin (57.2 s).*

Damage to the skin begins after the basal layer temperature increases above 44°C, nominally 35.1 s in Fig. 5. The heat source is removed 2.6 s later. The basal layer temperature continues to rise for an additional 2.6 s, before cooling below 44°C over a 16.9-s period. Using the Macor TSSS data, a breakdown of the amount of damage caused to the skin during the heating and post heating stages of this RPP test shows essentially all of the damage occurs after the removal of the heat source [4,5].

The Henriques Burn Integral can also be used to assess the time to pain. Pain will be felt when the thermal damage factor, Ω - the value of the integral in Eq 1, equals 0.53. In Fig. 5, this value for Ω is not reached until after the heat source is removed and the skin begins to cool. Consequently, by the time the wearer begins to feel pain, a 2nd degree burn is imminent unless the wearer can remove the clothing and immediately cool the skin. In these RPP tests, the time between when the wearer feels pain and when a 2nd degree burn will occur is between 5 and 10 seconds for the Nomex and PBI outer shell materials analyzed.

This type of analysis was completed on the ensemble samples as well, using the Henriques Burn Damage Integral heat flux information from the Macor TSSS. Even more of the damage was caused during the post heating stage of the ensemble evaluations, as more energy was stored within the clothing sample during the exposure. For each of the ensembles evaluated with the Macor skin simulant, less than 1 % of the thermal damage was caused during the actual heating stage of the test, which lasted approximately 30 s [4,5].

Thermal Properties

Although it is not part of current performance test methods, evaluating the thermal properties of protective clothing components is useful for developing engineering models of the clothing and for estimating the conductive portion of total heat loss [11,20]. A Thermal Property Test Fixture (TPTF) was developed to expose the clothing to low heating levels for non-destructive evaluation of the clothing's properties. The heat flux and temperature data from the TPTF was then input into a parameter estimation code, PROP1D [2], to assess the materials' thermal properties [4,7,8].

Several outer shell materials were evaluated using this test apparatus. Table 1 indicates thermal properties determined for various outer shell materials (Nomex and PBI), two fire retardant fabrics used by the military (Southern Mills Green and Navy materials), as well as one representative thermal liner and one moisture barrier material. The estimates compare well with data presented in the literature [20]. These thermal properties were then used to estimate the time to 2nd degree burn using a numerical model developed by Gagnon [4].

Several ensembles were also evaluated in the TPTF to determine their overall thermal properties. The ensembles used Nomex 7.5, Nomex 6.0, and PBI 7.5 oz/sq yd fabrics as the outer shell materials, coupled with a moisture barrier and the thermal liner. The thermal liner material was located behind the moisture barrier, closest to the skin. The thermal liner serves as the primary insulating layer. Consequently, the effective thermal properties of the three ensembles were similar.

TABLE 1—*Dry material thermal property estimates.*

Material (Weight oz/sq yd)	Thermal Conductivity (W/m-K)	Volumetric Heat Capacity (J/m^3-K)
Nomex (7.5)	0.053	1.56×10^6
Nomex (6.0)	0.043	1.56×10^6
Nomex (5.5)	0.043	1.53×10^6
PBI (7.5)	0.051	1.57×10^6
PBI (6.0)	0.045	1.58×10^6
Southern Mills Green (8.5)	0.044	1.63×10^6
Navy (6.8)	0.043	1.62×10^6
Aralite Thermal Liner	0.034	7.46×10^5
Crosstech Moisture Barrier	0.041	9.72×10^5

The clothing materials may become damp or wet during actual use due to water spray or perspiration of the user. As Lawson [10] notes, small amounts of moisture may be beneficial. Steady state methods, such as a Guarded Hot Plate, cannot be used to measure the thermal conductivity during drying. Dynamic experiments with the TPTF were used to evaluate the thermal properties of clothing materials with various amounts of moisture added to simulate actual use conditions.

The addition of moisture to the clothing materials increases both the thermal conductivity and the volumetric heat capacity; this leads to an increase in heat transfer as compared with dry materials during similar test exposure conditions [4]. Extensive details on the thermal property measurements made during this project are beyond the scope of this paper; these results are available in the SBIR Summary Reports [7,8] and in a WPI Master's Thesis [4]. The reports contain information on thermal property measurements of:

- dry fabrics and ensembles
- wetted fabrics and ensembles
- fabrics under compression
- jackets using a hot surface, sudden contact technique

Summary and Conclusions

The current work has shown that burn time estimates provided by the current test methods appear to be significantly higher – the measured range was 23–34 % – than the actual exposure time required to produce a 2nd degree burn. More accurate measurements of thermal protection can be obtained by using a higher fidelity heat source and thermal skin simulant sensors. Thermal property measurements are useful for developing engineering models to predict protective clothing performance under a variety of use conditions, such as the long-duration, low-heat flux evaluations conducted at NIST-BFRL.

Suggested Changes

1. Replace the lamp/burner combination in the TPP procedure (and the lamps in the RPP) with a higher fidelity radiant heat source to more closely simulate heat transfer from a fire and to help reduce uncertainty. A Cone Calorimeter heater is one possible choice [1].
2. Replace the Copper Calorimeter with a Thermal Skin Simulant Sensor (TSSS), or modify the copper calorimeter design to more closely replicate the skin's surface temperature response.
3. Use a Two Step Analysis Method to Estimate Time to a Second-Degree Burn and Total Burn Damage. Calculate a heat flux history from temperatures measured with the calorimeter or a TSSS. Then, estimate the time to a 2nd degree burn and total burn damage using Henriques Burn Damage Integral or a Skin Model [9].
4. Although not directly part of the current thermal protection test methods, suggest Obtaining Thermal Properties For Modeling Clothing Performance. Use non-destructive test methods for determining material properties of individual layers or ensembles, including effects of moisture or compression. These are described in Keltner, et al. [8] and Gagnon [4]. The models can be used to predict protective performance during the wide variety of exposures experienced by fire fighters.

Acknowledgments

The author gratefully acknowledges the substantial contributions of many people. Randy Lawson, NIST-BFRL, was the Technical Project Officer for this SBIR project and provided a wealth of information to help me learn. Thanks to Professors Nicholas Dembsey and Edward Clougherty from WPI, plus two of their graduate students – Dana Haagensen, who made major contributions across the board in Phase I, and Brian Gagnon, who generated detailed information during his Master's Thesis in Phase II, which is referenced heavily in this paper. Dr. Roger Parry, Dupont, provided much insight into techniques for analyzing test data and also provided demonstrations of TPP and flash fire test procedures.

PostScript

With the exception of a few proprietary items, all of the information developed in the project is available without fee by contacting the author.

References

[1] ASTM Standard E 1354, "Standard Test Methods for Heat and Visible Smoke Release Rates for Materials and Products Using an Oxygen Consumption Calorimeter," ASTM International, West Conshohocken, PA.

[2] Beck, J. V. and McMasters, R. L., "Users Manual for Prop1D – Program for Estimating Thermal Properties from Transient Temperature and Heat Flux Measurements," Version 7.0, and, " Users Manual for IHCP1D – Program for Calculating Surface Heat Fluxes from Transient Temperatures Inside Solids," Beck Engineering Consultants Co., Okemos, MI.

[3] Duck, F. A., *Physical Properties of Tissue – A Comprehensive Reference Book*, Academic Press, 1990, ISBN 0-12-222800-6.

[4] Gagnon, B. D., "Evaluation of New Test Methods for Fire Fighting Clothing," Master's Thesis, Worcester Polytechnic Institute, May 2000.

[5] Gagnon, B. D., Dembsey, N. A., and Keltner, N. R., "Preliminary Results from New Methods for Evaluation of Fire Fighting Clothing," *Fire and Materials Conf.*, San Francisco, 2001.

[6] Henriques, F. C., Jr., "Studies of Thermal Injury V. The Predictability and the Significance of Thermally Induced Rate Processes Leading to Irreversible Epidermal Injury," *Archives of Pathology*, Vol. 43, 1947, pp. 489–502.

[7] Keltner, N. R., et al., "New Methods for Evaluating Thermal Performance of Protective Clothing for Fire Fighters – Phase I," Ktech Corporation, Technical Report TR98-01, January 1998.

[8] Keltner, N. R., et al., "New Methods for Evaluating Thermal Performance of Protective Clothing for Fire Fighters – Phase II," Ktech Corporation, Technical Report TR00-15, November 2000.

[9] Knox, III, F. S, et al., "User's Manual for BRNSIM/BURNSIM – a Burn Hazard Assessment Model," Wright-Patterson AFB, OH, March 1992.

[10] Lawson, J. R., "Thermal Performance and Limitations of Bunker Gear," *Fire Engineering*, August 1998.

[11] Mell, W. E. and Lawson, J. R., "A Heat Transfer Model for Fire Fighter's Protective Clothing," U.S. Department of Commerce, NISTIR 99-6299, August 1999.

[12] NFPA 1971, *Protective Clothing Ensemble for Structural Fire Fighting*, 1997 ed., National Fire Protection Association, Quincy, MA, 1996.
[13] NFPA 1977, *Protective Clothing and Equipment for Wildland Fire Fighting*, 1993 ed., National Fire Protection Association, Quincy, MA, 1996.
[14] Quintiere, James, "Radiative Characteristics of Fire Fighter's Coat Fabrics," *Fire Technology*, Vol. 10, No. 2, May 1974, pp. 153–161.
[15] Reed, R. P., "Convolution & Deconvolution in Measurement and Control," *Measurements & Control*, Eight Part Professional Course, April 1997–September 1998.
[16] SFPE Task Group on Engineering Practices, "Predicting 1st and 2nd Degree Skin Burns from Thermal Radiation," *Society of Fire Protection Engineers*, Bethesda, MD, 2000.
[17] Stoll, A. M. and Chianta, M. A., "Method and Rating System for Evaluation of Thermal Protection," *Aerospace Medicine*, Vol. 40, 1968, pp. 1232–1238.
[18] Stoll, A. M. and Greene L. C., "Relationship Between Pain and Tissue Damage Due to Thermal Radiation," *Journal of Applied Physiology*, Vol. 14, 1959, pp. 373–382.
[19] Stoll, A. M. and Chianta, M. A., "Heat Transfer through Fabrics as Related to Thermal Injury," *Transactions-New York Academy of Sciences*, Vol. 33, No. 7, 1971, pp. 649–670.
[20] Torvi, D. A. and Dale, J. D., "Numerical Models for Use in the Design of Protective Clothing for Firefighters," ICFRE 3, Chicago, 1999.

Journal of ASTM International, January 2005, Vol. 2, No. 1
Paper ID JAI12115
Available online at www.astm.org

David A. Torvi[1]

Effects of Thermal Properties on Skin Burn Predictions in Longer Duration Protective Clothing Tests

ABSTRACT: Different methods of predicting skin burns during bench top and full-scale tests of protective clothing are compared. A skin heat transfer and burn prediction model is also used to determine the effects of thermal properties of skin and blood flow rates on first, second, and third degree skin burn predictions for exposures used in longer duration tests of single and multiple-layer protective fabrics. Epidermis thickness can have a large effect on second degree burn predictions, while the dermis thickness can have a very large effect on third degree burn predictions. Subcutaneous thickness does not significantly affect burn predictions, while blood flow can affect third degree burn predictions. Results also indicate that care must be taken when analyzing burn predictions, as relatively small differences in skin temperatures can produce deceptively large differences in burn predictions when Henriques' burn integral is used.

KEYWORDS: protective clothing, skin burn, firefighter, bioheat transfer, turnout gear, thermal protective performance test, radiant protective performance test

Nomenclature

A	surface area (m^2)
c	specific heat (J/kg·°C)
G	blood perfusion rate ($m^3/s/m^3$ tissue)
K	heat loss coefficient for test sensor (W/m^2·°C)
k	thermal conductivity (W/m·°C)
L	thickness (m)
m	mass (kg)
P	pre-exponential factor (s^{-1})
q"	heat flux (W/m^2)
R	ideal gas constant (J/mol·K)
T	temperature (°C, K)
t	time (s)
x	depth (m)

Greek symbols

ΔE	activation energy (J/mol)
Ω	Henriques' burn integral value (dimensionless)
ρ	density (kg/m^3)

Manuscript received 13 November 2003; accepted for publication 8 July 2004; published January 2005. Presented at ASTM Symposium on Performance of Protective Clothing: Global Needs and Emerging Markets: 8th Symposium on 13–15 January 2004 in Tampa, FL.

[1] Ph.D. and Associate Professor, Dept. of Mechanical Engineering, University of Saskatchewan, Saskatoon, SK S7N 5A9.

Subscripts

b blood
c core
i initial

Introduction

Standard tests used to evaluate the thermal protection offered by fabrics and garments typically provide users of these test methods with an estimate of times required for skin burn damage to occur. Bench top tests, such as the TPP (the former ASTM Standard Test Method for Thermal Protective Performance of Materials for Clothing by Open-Flame Method (D 4108)) and the RPP (ASTM Standard Test Method for Radiant Protective Performance of Flame Resistant Clothing Materials (F 1939)) tests, use the Stoll second-degree burn criterion [1]. Full-scale mannequin tests, such as the ASTM Standard Test Method for Evaluation of Flame Resistant Clothing for Protection Against Flash Fire Simulations Using an Instrumented Manikin (F 1930), use Henriques' burn integral [2]. Both of these techniques provide an estimate of the time required to produce second degree burns, while Henriques' burn integral can also provide an estimate of the time required to produce third degree burns.

Previous research has demonstrated that for short duration tests, such as TPP tests of single-layer fabrics, both skin burn prediction techniques give similar results [3]. Research has also shown that it is not important to consider most physiological changes the body makes in response to a thermal insult when making skin burn predictions in short duration tests [4]. This is because the length of time required to produce skin burns (e.g., 6–7 s) is shorter than the length of time for physiological changes to occur (e.g., it takes approximately 20 s for the body to increase the rate of blood flow [5]).

Tests of multiple-layer fabrics, such as firefighter turnout gear, are considerably longer than single-layer fabric tests. For example, the NFPA Standard on Protective Ensemble for Structural Fire Fighting (NFPA 1971) contains a requirement that turnout gear should have a minimum TPP value of 35, which corresponds to a test of at least 17.5 s. Other tests are being developed which will evaluate the stored energy in a garment and the effects on human skin when this energy is released [3]. These tests could last more than one minute. These tests are important because numerical models of heat transfer in turnout gear have demonstrated that stored energy can have a significant effect on slow developing third-degree burns [6]. For longer duration exposures, physiological effects, such as increased blood flow or sweating, could begin to affect skin temperatures, and hence if and when skin burns occur. Therefore, it is important to consider these physiological effects when making skin burn predictions using data from longer duration protective clothing tests.

Another use of skin heat transfer and burn models is to predict skin damage in larger numerical models of fabric-covered skin, such as models of heat transfer in turnout gear [6]. Models are being developed to help firefighters and manufacturers understand the physics of heat and moisture transfer during relatively long duration exposures to lower level heat fluxes. Skin burn models used in conjunction with these fabric heat and mass transfer models may need to include these additional physiological effects. Another application of skin burn models is in the field of fire protection engineering, where hazard analyses are conducted to determine the expected effect of fire on people. While traditional methods of predicting skin burns rely on correlations that are based on data from the atomic explosions in Hiroshima and Nagasaki [7], a new method has recently been developed that incorporates a skin heat transfer model, Henriques' burn integral, and mortality data from burn wards [8].

As a first step toward the development of these more sophisticated skin heat transfer and burn models, a parametric study has been performed in order to determine the effects of thermal properties and blood flow on skin temperatures and predicted times required to produce second and third degree burns. This study is similar to studies conducted by other investigators for various boundary conditions, including: three second exposures to constant heat fluxes of between 24 and 166 kW/m^2 [4], three second exposures to time varying heat fluxes with peak values of between 20 and 80 kW/m^2 [9], 15 s exposures with a constant skin surface temperature of 90°C [10], and exposures that use a convective boundary condition to simulate scald injury [11].

In this paper, burn predictions made using the Stoll criterion and Henriques' burn integral will be compared for longer duration exposures indicative of those in TPP and RPP tests of single and multiple-layer fabrics. Parametric studies will be performed to determine the effects of changes in selected thermal properties on skin burn predictions for these exposures. As the durations of these tests range from 60–105 s, these parametric studies will provide information on the importance of thermal properties under conditions that are considerably different from those used in the parametric studies discussed above. In addition, the results of these parametric studies will also help to provide guidance as to where effort should be concentrated when developing new skin heat transfer and burn models for situations where additional physiological effects, such as increases in blood flow and sweat, should be incorporated.

It should be noted that besides choosing appropriate values of thermal properties, there are many other issues that must be considered when specifying the method used to predict skin burn injury in protective clothing tests or in other applications. While this study only considered the Stoll criterion and Henriques' Burn Integral, there are other techniques that have been developed to make skin burn predictions. The nature of the exposure for which skin burn predictions are being made must be considered, as different techniques have been validated using data from exposures to different modes of heat transfer, and different magnitudes of temperatures and heat fluxes. Clinically, third degree burns are more difficult to diagnose than second degree burns, which has implications on the relative confidence which can be placed in second and third degree burn predictions. Discussion of these and other important issues is outside of the scope of this paper, but it can be found in other references (e.g., [12,13]).

Heat Flux Histories

In order to compare skin burn predictions under a variety of conditions, heat flux histories were determined using copper disk test sensor data from the following three tests of single and multiple-layer fabrics conducted using the ASTM D 4108 (TPP) and ASTM F 1939 (RPP) tests (Fig. 1):

- A test of a single layer inherently flame resistant fabric using the TPP test apparatus (10 s exposure to a nominal heat flux of 80 kW/m^2, followed by cooling for 50 s);
- A test of a firefighters' ensemble (outer shell, moisture barrier, and thermal liner) using the TPP test apparatus (20 s exposure to a nominal heat flux of 80 kW/m^2, followed by cooling for 40 s); and
- A test of a firefighters' ensemble (outer shell, moisture barrier, and thermal liner) using the RPP test apparatus (105 s exposure to a nominal heat flux of 20 kW/m^2).

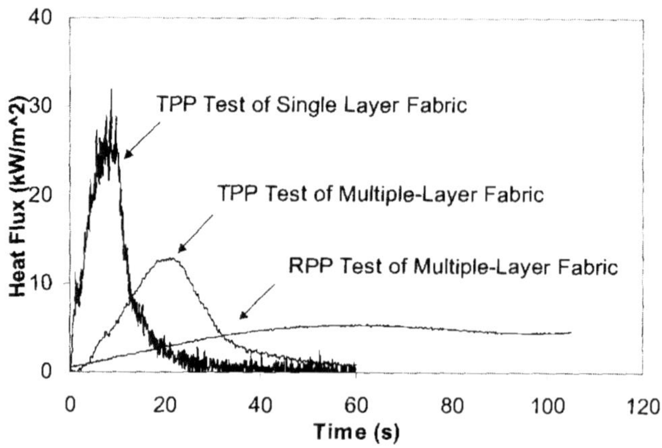

FIG. 1—*Heat flux histories taken from three bench top fabric tests and used to compare skin burn predictions.*

Heat fluxes were determined from test sensor temperatures using a computer program that is based on a lumped heat capacity analysis of the test sensor [3]:

$$q''(t) = \frac{mc}{A}\frac{dT}{dt} + K\big(T(t) - T(0)\big) \tag{1}$$

The five-point linear least squares method was used to solve Eq 1, as it was previously demonstrated that this method helps to reduce fluctuations in the calculated heat flux histories [3].

Skin Burn Prediction Techniques Used

Stoll Second Degree Burn Criterion

While the Stoll second degree burn criterion is strictly only valid for an exposure to a constant heat flux level [1], it is used in many bench top tests to evaluate the thermal protection offered by candidate fabrics. The energy absorbed by the copper calorimeter in a test can be compared with Stoll and Chianta's data for the total amount of energy that must be absorbed by skin to produce a second degree burn for a given length of exposure. In this study, and in bench top tests, this is done by comparing the increase in copper calorimeter temperature at any time with a critical temperature increase that is based on Stoll and Chianta's data.

Skin Heat Transfer Model and Henriques' Burn Integral

The skin heat transfer model uses the Pennes' bioheat transfer equation [14] to predict temperatures within the skin:

$$\rho c\frac{\partial T}{\partial t} = k\frac{\partial^2 T}{\partial x^2} - G(\rho c)_b(T - T_c) \tag{2}$$

This differential equation is solved subject to the following initial condition:

$$T(x,t=0) = T_i(x) \tag{3}$$

where $T_i(x)$ is a quadratic initial temperature between the skin surface temperature (32.5°C) and the body core temperature (37°C). This temperature distribution is consistent with experimental data [14].

The following boundary conditions are used. First, the temperature of the base of the subcutaneous layer is assumed to remain at the body core temperature, which is assumed to remain constant during the time of interest:

$$T(x = L,t) = T_c(x) \tag{4}$$

The heat flux history ($q''(t)$) on the surface of the skin is incorporated within the other boundary condition:

$$k\left(\frac{\partial T}{\partial x}\right) + q''(t) = 0 \qquad (x = 0,t) \tag{5}$$

Once the temperatures within the skin are determined, Henriques' burn integral [2] is used to estimate the times required to first, second, and third degree burn:

$$\Omega = \int_0^t P\,exp\left(-\frac{\Delta E}{RT}\right)dt \tag{6}$$

In order to predict first and second degree burns, the temperature (T) of the base of the epidermis (basal layer) is used in Eq 6, and the equation is integrated over the time (t) for which the basal layer temperature remains above 44°C, which is the temperature at which thermal damage begins. First degree burns are said to occur when the value of Ω reaches 0.53 in Eq 6, while second degree burns are predicted to occur when the value of Ω reaches 1.0. To predict third degree burns, the temperature (T) of the base of the dermis is used in Eq 6, and the equation is integrated over the time (t) for which this temperature remains above 44°C. Third degree burns are said to occur when the value of Ω reaches 1.0.

Various thermal properties of the skin and the blood appear in Eqs 2–6. The values of these properties used in the heat transfer model are shown in Table 1. Details on the sources of these thermal properties and the variation in values published in the literature can be found in [4].

Results and Discussion

Comparison of Skin Burn Prediction Techniques

The Stoll second degree burn criterion and Henriques' burn integral were used to predict the times required to produce first, second, and third degree burn for the three heat flux histories (Table 2). Predictions made using the Stoll criterion and Henriques' burn integral for TPP tests of two single-layer fabrics also were compared in an earlier paper [3]. In this study, differences between predicted times required to produce second degree burns ranged between 9 % and 11 % for the two heat flux histories taken from TPP and RPP tests of turnout gear. These differences are consistent with the earlier results for single-layer TPP tests. As mentioned earlier, the Stoll criterion is strictly only valid for square wave heat flux histories. Figure 1 demonstrates that the heat flux histories in longer duration tests of turnout gear are not square waves. However, the results in Table 2 indicate that if only second degree burn estimates are of interest, the Stoll criterion can continue to be used in tests of turnout gear that are similar to the tests conducted in this study.

TABLE 1—*Thermal properties used in the skin heat transfer model [4].*

Thermal Property	Value	
Specific Heat (c), J/kg·°C		
- epidermis	3598	
- dermis	3222	
- subcutaneous	2760	
- blood	3770	
Blood perfusion rate (G), $m^3/s/m^3$ tissue		
- epidermis	0	
- dermis	0.00125	
- subcutaneous	0.00125	
Thermal conductivity (k), W/m·°C		
- epidermis	0.255	
- dermis	0.523	
- subcutaneous	0.167	
Thickness (L), m		
- epidermis	80×10^{-6}	
- dermis	0.002	
- subcutaneous	0.01	
Body core temperature (T_c), °C	37	
Skin surface temperature ($T_{x=0}$), °C	32.5	
Density (ρ), kg/m^3		
- epidermis	1200	
- dermis	1200	
- subcutaneous	1000	
- blood	1060	
Pre-exponential factor (P), s^{-1}		
- basal layer	2.185×10^{124}	$44 \le T < 50°C$
	1.823×10^{51}	$T \ge 50°C$
- base of dermis	4.32×10^{64}	$44 \le T < 50°C$
	9.39×10^{104}	$T \ge 50°C$
Ratio of activation energy to ideal gas constant (ΔE/R), K		
- basal layer	93 534.9	$44 \le T < 50°C$
	39 109.8	$T \ge 50°C$
- base of dermis	50 000	$44 \le T < 50°C$
	80 000	$T \ge 50°C$

TABLE 2—*Comparison of skin burn predictions for bench top protective clothing tests utilizing three different heat flux histories.*

Fabric Specimen and Bench Top Test Exposure	Predicted Time to Produce 2nd Degree Burn Calculated Using Stoll Criterion, s	Predicted Times to Produce Burns Calculated Using Skin Heat Transfer Model and Henriques' Burn Integral		
		1st Degree Burn, s	2nd Degree Burn, s	3rd Degree Burn, s
Single Layer – 80 kW/m^2 for 10 s	6.38	5.66	5.88	32.63
Multiple Layer – 80 kW/m^2 for 20 s	19.97	17.70	18.33	57.05
Multiple Layer – 20 kW/m^2 for 105 s	51.2	43.4	46.0	81.3

Effects of Skin Layer Thickness Values on Burn Predictions

Previous research investigated the effects of the values of the thermal properties in Eqs 2–6 on predicted skin temperatures and burns [4] for a series of constant heat flux exposures from 24–166 kW/m^2. It was found that variations in thermal properties had minimal effects on second degree burn predictions, but larger effects on third degree burn predictions. As the previous study did not investigate the effects of the thickness of individual layers of skin, it was decided to investigate these effects in this research.

Human skin consists of three layers: the epidermis, the dermis, and the subcutaneous layer. The thickness of each of these layers varies from person to person and from site to site on an individual. A search of the literature revealed the range of values reported for the thickness of these three layers, as well as the range of values used in skin heat transfer models [15]. If palms and soles are neglected, the mean thickness of the epidermis in most regions of the body ranges between 20–80 μm. Most skin heat transfer models use values between 75–120 μm. The thickness of the dermis ranges between 0.45–2.6 mm, while most models are between 0.8–2.0 mm. Most models use a value between 0.3–1.0 cm for the thickness of the subcutaneous layer.

As an example of the magnitude of the difference in burn predictions due to differences in the values used for the thickness of individual layers, skin burn predictions were made using the epidermis, dermis, and subcutaneous layer thickness values listed in Table 1 (80 μm, 2.0 mm, and 1.0 cm, respectively) and ASTM F 1930 (50 μm, 1.5 mm, and 1.0 cm). Values listed in Table 1 were used for all other thermal properties. First, second, and third degree burn predictions for each of the three heat flux histories are shown in Table 3. First and second degree burn predictions were within 2–3 % of one another for the two TPP tests and within 5 % of one another for the RPP test. There were significant differences in the predicted times to produce third degree burn in all tests. These differences were larger in the two TPP tests (54–67 %) than in the RPP test (19 %).

To investigate the effects of the thickness of each layer, the epidermis thickness was varied between 50–150 μm (Table 4), the dermis thickness was varied between 1.0–2.5 mm (Table 5), and the subcutaneous layer thickness was varied between 0.5–2.0 cm (Table 6). Values listed in Table 1 were used for all other thermal properties.

TABLE 3—*Comparison of skin burn predictions for bench top protective clothing tests using skin heat transfer model and skin layer thickness values listed in ASTM F 1930 and Table 1.*

Fabric Specimen and Bench Top Test Exposure	Predicted Times to Produce Burns Calculated Using Skin Heat Transfer Model and Henriques' Burn Integral					
	1st Degree Burn, s		2nd Degree Burn, s		3rd Degree Burn, s	
	ASTM F 1930	Table 1	ASTM F 1930	Table 1	ASTM F 1930	Table 1
Single Layer – 80 kW/m^2 for 10 s	5.50	5.66	5.72	5.88	16.26	32.63
Multiple Layer – 80 kW/m^2 for 20 s	17.33	17.70	17.96	18.33	32.93	57.05
Multiple Layer – 20 kW/m^2 for 105 s	41.4	43.4	43.7	46.0	67.3	81.3

TABLE 4—*Predicted times to produce skin burns calculated using skin heat transfer model and Henriques' burn integral for various values of epidermis layer thickness.*

Heat Flux History	Epidermis Thickness, μm			
	50	80	100	150
First Degree Burns				
Single Layer – 80 kW/m² for 10 s	5.39	5.66	5.88	6.21
Multiple Layer – 80 kW/m² for 20 s	17.33	17.70	18.59	19.09
Multiple Layer – 20 kW/m² for 105 s	42.8	43.4	49.3	50.5
Second Degree Burns				
Single Layer – 80 kW/m² for 10 s	5.72	5.88	5.94	6.38
Multiple Layer – 80 kW/m² for 20 s	18.08	18.33	18.96	19.59
Multiple Layer – 20 kW/m² for 105 s	45.3	46.0	50.9	52.1
Third Degree Burns				
Single Layer – 80 kW/m² for 10 s	30.49	32.63	34.28	39.77
Multiple Layer – 80 kW/m² for 20 s	54.16	57.05	59.19	no burn
Multiple Layer – 20 kW/m² for 105 s	80.4	81.3	81.9	83.5

TABLE 5—*Predicted times to produce skin burns calculated using skin heat transfer model and Henriques' burn integral for various values of dermis layer thickness.*

Heat Flux History	Dermis Thickness, mm			
	1.0	1.5	2.0	2.5
First Degree Burns				
Single Layer – 80 kW/m² for 10 s	5.66	5.66	5.66	5.66
Multiple Layer – 80 kW/m² for 20 s	17.33	17.58	17.70	17.70
Multiple Layer – 20 kW/m² for 105 s	39.4	42.0	43.4	44.2
Second Degree Burns				
Single Layer – 80 kW/m² for 10 s	5.88	5.88	5.88	5.88
Multiple Layer – 80 kW/m² for 20 s	17.96	18.33	18.33	18.33
Multiple Layer – 20 kW/m² for 105 s	41.6	44.3	46.0	46.9
Third Degree Burns				
Single Layer – 80 kW/m² for 10 s	11.32	16.76	32.63	no burn
Multiple Layer – 80 kW/m² for 20 s	25.38	33.69	57.05	no burn
Multiple Layer – 20 kW/m² for 105 s	55.9	68.1	81.3	96.0

TABLE 6—*Predicted times to produce skin burns calculated using skin heat transfer model and Henriques' burn integral for various values of subcutaneous layer thickness.*

Heat Flux History	Subcutaneous Layer Thickness, cm			
	0.5	1.0	1.5	2.0
First Degree Burns				
Single Layer – 80 kW/m^2 for 10 s	5.55	5.66	5.66	5.66
Multiple Layer – 80 kW/m^2 for 20 s	17.45	17.70	17.83	17.83
Multiple Layer – 20 kW/m^2 for 105 s	42.2	43.4	44.1	44.7
Second Degree Burns				
Single Layer – 80 kW/m^2 for 10 s	5.88	5.88	5.94	5.94
Multiple Layer – 80 kW/m^2 for 20 s	18.08	18.33	18.46	18.59
Multiple Layer – 20 kW/m^2 for 105 s	44.7	46.0	46.7	47.3
Third Degree Burns				
Single Layer – 80 kW/m^2 for 10 s	29.61	32.63	34.28	37.30
Multiple Layer – 80 kW/m^2 for 20 s	51.14	57.05	no burn	no burn
Multiple Layer – 20 kW/m^2 for 105 s	79.3	81.3	82.4	83.5

The thickness of the epidermis had a larger effect on first and second degree burn predictions for the RPP test than for the TPP tests. For example, doubling the thickness from 50 to 100 μm increased the time to second degree burn by 12 % for the RPP test and by 4–5 % for the two TPP tests. On the other hand, increasing the epidermis thickness had a larger effect on third degree burn predictions for the TPP tests than the RPP test. Doubling the thickness from 50 to 100 μm increased the time to third degree burn by 9–12 % for the two TPP tests and by 2 % for the RPP test.

Further examination of these results, however, reveals how differences in skin burn predictions can be deceptive due to the nonlinear nature of Henriques' burn integral (Eq 6). Figure 2 shows the basal layer temperature histories predicted for the TPP and RPP tests of turnout gear, using epidermis thickness values of 50 and 100 μm. There is an appreciable difference in the basal layer temperatures, and hence the second degree burn predictions for the TPP test. However, the basal layer temperatures for the RPP test are quite close to one another, yet the predicted time to second degree burn increased from 45.3 to 50.9 s when the epidermis thickness was doubled. This is due to the fact that the rate of temperature rise in the basal layer for this exposure is so slow that any slight change of temperature can cause a significant difference in the time required for the value of Henriques' burn integral to reach 1.0.

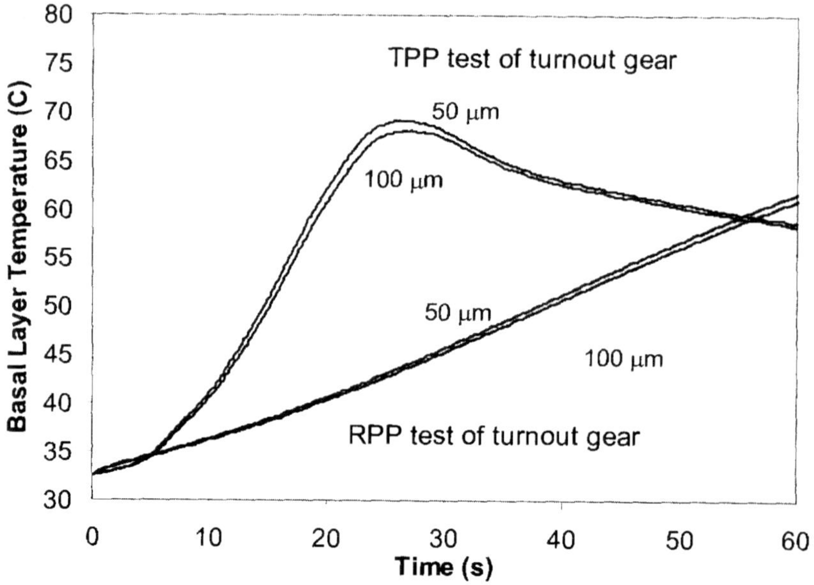

FIG. 2—*Basal layer temperatures predicted by the skin heat transfer model for TPP and RPP tests of turnout gear and two values of epidermis thickness.*

The thickness of the dermis had little or no effect on first and second degree burn predictions, but it had a large effect on third degree predictions for all tests. The effect of the dermis thickness was much larger in TPP tests than in the RPP test. For example, increasing the thickness of the dermis from 1.5 to 2.0 mm (a 33 % increase) increased the time required to produce third degree burns by 51–64 % for the TPP tests and by 18 % for the RPP test. Further examination revealed that the dermis thickness has a significant effect on skin temperatures, and hence differences in skin burn predictions are not merely a result of the nonlinear nature of Henriques' burn integral.

The thickness of the subcutaneous layer had a very minor effect on first and second degree burn predictions. In most tests, increasing the subcutaneous layer thickness had a minor effect on third degree burns. Upon further review, differences in third degree burn predictions for the TPP test of turnout gear in this part of the study were found to be mainly due to the nonlinear nature of Henriques' burn integral.

Effect of Blood Flow on Skin Burn Predictions

In order to investigate the effects of blood flow on skin burn predictions for these three tests, skin burn predictions were compared for three rates of blood perfusion: 0, 0.00125, and 0.0025 $m^3/s/m^3$ tissue (Table 7). These rates represent no blood flow, normal blood flow, and elevated blood flow, which may occur in response to a longer duration exposure. Values listed in Table 1 were used for all other thermal properties.

TABLE 7—*Predicted times to produce skin burns calculated using skin heat transfer model and Henriques' burn integral for various blood perfusion rates.*

Heat Flux History	Blood Perfusion Rate, $m^3/s/m^3$ tissue		
	0.0	0.00125	0.0025
First Degree Burns			
Single Layer – 80 kW/m^2 for 10 s	5.66	5.66	5.66
Multiple Layer – 80 kW/m^2 for 20 s	17.70	17.70	17.70
Multiple Layer – 20 kW/m^2 for 105 s	43.4	43.4	43.5
Second Degree Burns			
Single Layer – 80 kW/m^2 for 10 s	5.88	5.88	5.88
Multiple Layer – 80 kW/m^2 for 20 s	18.33	18.33	18.33
Multiple Layer – 20 kW/m^2 for 105 s	45.9	46.0	46.1
Third Degree Burns			
Single Layer – 80 kW/m^2 for 10 s	30.43	32.63	36.14
Multiple Layer – 80 kW/m^2 for 20 s	52.65	57.05	no burn
Multiple Layer – 20 kW/m^2 for 105 s	79.7	81.3	83.1

Table 7 indicates that blood flow had practically no effect on first and second degree burn predictions for these three exposures. The effects of blood flow on third degree burn predictions were relatively minor for the tests of the turnout gear, but significant for the TPP test of the single-layer fabric. For example, doubling the blood perfusion rate to simulate physiological changes which the body could make in response to a high heat flux exposure increased the time to third degree burn by 2 % for the RPP test and 10 % for the TPP test of the single-layer fabric. However, once again some of the differences in this parametric study can be deceptive. For example, in the TPP test of the turnout gear, increasing the blood perfusion rate from 0.00125 to 0.0025 $m^3/s/m^3$ tissue decreased the temperature of the dermal base at 60 s from 54 to 53.3°C, which was enough to prevent third degree burns from being predicted for the higher perfusion rate.

Conclusions and Future Work

This work has demonstrated that second degree burn predictions made using the Stoll criterion and Henriques' burn integral are similar for TPP and RPP tests of single and multiple-layer fabrics. The thickness of individual skin layers and the blood perfusion rate affect skin burn predictions to various degrees. Epidermis thickness can have a large effect on second degree burn predictions, and dermis thickness can have a very large effect on third degree burn predictions. Subcutaneous thickness does not have a significant effect on either second or third degree burn predictions. Blood perfusion rate also can affect third degree burn predictions. Results also demonstrate that care must be exercised when investigating the effects of thermal properties on skin burn predictions, due to the nonlinear nature of Henriques' burn integral.

For some applications, it may be useful to continue to make skin heat transfer and burn prediction models more realistic by incorporating physiological effects, such as increased blood flow and sweat that occur during longer duration exposures. For example, Ng and Chua [11] discuss how the effects of sweating can be incorporated in a model by increasing the thermal conductivity of the skin. While the research in this paper demonstrates that increasing blood flow may not have as large of an effect on skin burn predictions as other thermal properties for some of the heat flux histories used in this study, the effect of sweating on thermal conductivity and other methods of incorporating physiological effects in burn models should continue to be investigated.

In the range of heat fluxes measured during these tests, it was found that burn predictions can be very sensitive to small changes in thermal properties within the model. Therefore, as additional tests are developed for longer duration exposures, it will be very important to explicitly state in test standards the values of thermal properties that must be included in skin burn models used to analyze data from these new tests.

Due to the nature of burn criteria (either a burn or no burn is predicted), it may also be desirable to establish an alternative method of ranking fabrics based on a time to achieve a specified test sensor temperature rise or a predicted temperature at a particular location in human skin. As such a method would not rely on the nonlinear Henriques' burn integral, small differences in predicted skin temperatures would not be magnified. This approach is already used in the ISO International Standard for Protective Clothing Against Heat and Flame – Determination of Heat Transmission on Exposure to Flame (ISO 9151), where fabrics are rated on the time to a specified increase in copper calorimeter temperature rather than the time required to exceed the Stoll second degree burn criterion.

Acknowledgments

The author wishes to thank graduate students Peter Thorpe and Todd Threlfall, who performed the TPP and RPP tests of the turnout gear specimens. The financial support of the Natural Sciences and Engineering Research Council of Canada is also gratefully acknowledged.

References

[1] Stoll, A. M. and Chianta, M. A., "Method and Rating System for Evaluation of Thermal Protection," *Aerospace Medicine*, Vol. 40, 1969, pp. 1232–1238.

[2] Henriques, F. C., Jr., "Studies of Thermal Injuries V. The Predictability and the Significance of Thermally Induced Rate Processes Leading to Irreversible Epidermal Injury," *Archives of Pathology*, Vol. 43, 1947, pp. 489–502.

[3] Torvi, D. A., Dale, J. D., Ackerman, M. Y., and Crown, E. M., "A Study of New and Existing Bench Top Tests for Evaluating Fabrics for Flash Fire Protective Clothing," *Performance of Protective Clothing: 6th Volume, ASTM STP 1273*, J. O. Stull and Arthur D. Schwope, Eds., ASTM International, West Conshohocken, PA, 1997, pp. 108–125.

[4] Torvi, D. A. and Dale, J. D., "A Finite Element Model of Skin Subjected to a Flash Fire," *ASME Journal of Biomechanical Engineering*, Vol. 116, 1994, pp. 250–255.

[5] Lipkin, M. and Hardy, J. D., "Measurement of Some Thermal Properties of Human Tissues," *Journal of Applied Physiology*, Vol. 7, 1954, pp. 212–217.

[6] Torvi, D. A. and Dale, J. D., "Numerical Models for Use in the Design of Protective Clothing for Firefighters," *Journal of Industrial Textiles*, Vol. 29, 2000, pp. 273–286.

[7] Tsao, C. K. and Perry, W. W., "Modifications to the Vulnerability Model: A Simulation System for Assessing Damage Resulting from Marine Skills (VM4)," ADA 075 231, US Coast Guard NTIS Report No. CG–D–38–79, 1979.

[8] Torvi, D. A., Hadjisophocleous, G. V., and Hum, J., "A New Method for Estimating the Effects of Thermal Radiation from Fires on Building Occupants," *ASME 2000 International Mechanical Engineering Congress and Exposition (HTD–Vol. 366–5)*, 2000, pp. 65–72.

[9] Halin, J., Mulligan, J., Hamouda, H., Barker, R., and Soroka, A., "A Sensitivity Study of a Variable-Property Skin Burn Model Considering the Variability of Water Content," *Performance of Protective Clothing: Fifth Volume, ASTM STP 1237*, J. S. Johnson and S.

Z. Mansdorf, Eds., ASTM International, West Conshohocken, PA, 1996, pp. 592–606.

[10] Jiang, S. C., Ma, N., Li, H. J., and Zhang, X. X., "Effects of Thermal Properties and Geometrical Dimensions on Skin Burn Injuries," *Burns*, Vol. 28, 2002, pp. 713–717.

[11] Ng, E. Y-K. and Chua, L. T., "Prediction of Skin Burn Injury. Part 2: Parametric and Sensitivity Analysis," *Proceedings of the Institution of Mechanical Engineers. Part H: Journal of Engineering in Medicine*, Vol. 216, 2002, pp. 171–183.

[12] Diller, K. R. and Pearce, J. A., "Issues in Modeling Thermal Alterations in Tissues," *Annals of the New York Academy of Sciences*, Vol. 888, 1999, pp. 153–164.

[13] Diller, K. R., "Modeling of Bioheat Transfer Processes at High and Low Temperatures," *Advances in Heat Transfer*, Vol. 22, 1992, pp. 157–357.

[14] Pennes, H. H., "Analysis of Tissue and Arterial Blood Temperatures in Resting Human Forearm," *Journal of Applied Physiology*, Vol. 1, 1948, pp. 93–122.

[15] Torvi, D. A., "A Finite Element Model of Heat Transfer in Skin Subjected to a Flash Fire," M.Sc. Thesis, University of Alberta, Edmonton, AB, 1992.

Journal of ASTM International, July/August 2004, Vol. 1, No. 7
Paper ID JAI12116
Available online at www.astm.org

Guowen Song,[1] *R. Barker,*[2] *D. Thompson,*[2] *and R. Grimes*[2]

Effects of Simulated Flash Fire and Variations in Skin Model on Manikin Fire Test

ABSTRACT: An established numerical model of a manikin fire test, which has the capability of predicting heat transfer through thermally protective clothing exposed to an intense heat environment, is described in this paper. The model considers the fire characteristics simulated in a manikin chamber as well as the insulating air layers between protective garments and the skin surface. The numerical model is applied to analyze the effects of simulated flash fire and variations in a skin model on a manikin test. The study demonstrates that the heat flux measured by 122 thermal sensors over the surface of the manikin exhibits a bell-shaped Gaussian distribution for a short duration in calibration burn. A series of flash fire data with different distributions was generated statistically, and the effects on burn predictions were investigated. The results suggest that the fire distribution affects the burn predictions for 4 s of exposure. The effects of initial temperature distribution, thermal properties, as well as involvement of blood perfusion in a skin model on burn predictions are also discussed. The model predictions demonstrate that the initial temperature distribution in a skin model has a large effect on burn predictions for a one-layer garment exposed to short duration flash fire conditions.

KEYWORDS: skin model, flash fire, manikin fire test, fire distribution, thermal protective performance, temperature distribution, blood perfusion

Introduction

An instrumented manikin fire testing system is the most realistic laboratory assessment of the thermal protective performance of garments or ensembles. This manikin fire test system can control heat flux; flame distribution; duration; garment size, fit, and style; as well as the physical constitution of the wearer. Environmental conditions can be assessed in addition to the garment materials. This life-size thermal burn evaluation system is the fastest, most reproducible, and most advanced measurement system in the world today.

Figure 1 illustrates the basic elements of the Pyroman® [1] fire testing and burn evaluation processes. Laboratory flash fire is generated in the Pyroman® burn chamber by eight propane-burning torches. The heat from the fire transfers through the fabric and through the air gap between the protective garment and the manikin body. The temperatures are registered by 122 heat flux sensors on the manikin body. Predictions of skin burn injury resulting from the flash fire exposure are made using a skin model. The test procedure, data acquisition, result calculations, and test report preparation are performed with computer hardware and software programs.

Manuscript received 18 November 2003; accepted for publication 13 May 2004; published July/August 2004. Presented at ASTM Symposium on Performance of Protective Clothing: Global Needs and Emerging Markets: 8th Symposium on 13 January 2004 in Tampa, FL.

[1] Research Associate, Center for Research on Textile Protection and Comfort, College of Textiles, North Carolina State University, Raleigh, NC.
[2] Professor, Associate Director, and Special Project Director, respectively, College of Textiles, North Carolina State University, Raleigh, NC.

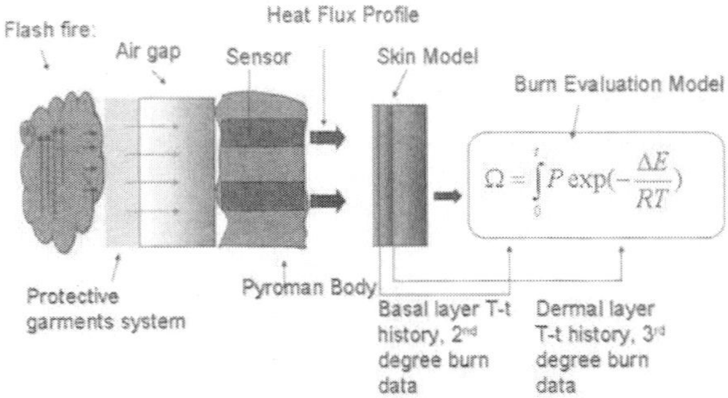

FIG. 1—*Heat transfer in instrumented manikin fire testing system.*

The flash fire generated in the Pyroman® chamber is by turbulent jet flames generated by eight propane torches around the manikin body. In the calibration process, the simulated flash fire is adjusted to produce an overall heat flux of 2.0 cal/cm^2·s for 4 s of exposure. The overall heat flux is an average heat flux measured by the 122 thermal sensors distributed over the manikin. The measured heat fluxes by the 122 sensors are not uniform because of the dynamic nature of the flame column surrounding the manikin and the complexity of human body geometry. The measured heat fluxes are then applied to a skin model programmed in the computer to calculate the temperature-time history in the skin layers. With this temperature-time history, burn injury predictions are made based on Henriques Burn Integral [2].

The literature proposes different skin models for human skin exposed to thermal hazards [3–5]. They differ in layer structure, layer thermal properties, initial temperature distribution, and blood perfusion. The simulated flash fire in the manikin chamber can produce different fire distributions but maintain an average heat flux of 2.0 cal/cm^2·s. In this paper an established numerical model of the manikin fire testing was used to examine the effect of fire distribution and variations in the skin model on manikin burn predictions for one-layer garments.

Numerical Model of Manikin Fire Testing

The model used by this research considers the entire burning process based on 122 sensor locations (Fig. 2). Heat transfer through the fabric and air layers is computed in conduction, radiation, and convection modes. A heat transfer equation is applied in conjunction with a skin model to estimate the temperature profile in the basal and dermal layers of the skin. Our model assumes that convective heat transfer occurs only at the surface of the protective fabric. Radiative heat flux is assumed to penetrate the fabric only to a certain depth. Based on these assumptions, the energy balance in a fabric element is [6]:

$$\rho_{fab}(T)Cp_{fab}(T)\frac{\partial T}{\partial t} = \frac{\partial}{\partial x}\left(k_{fab}(T)\frac{\partial T}{\partial x}\right) - \gamma \cdot q_{rad}e^{-\gamma x} \tag{1}$$

where ρ_{fab} is the density of the fabric; Cp_{fab} is the specific heat; k_{fab} is the thermal conductivity; and γ is the extinction coefficient of the fabric determined from transmissivity and the fabric thickness. The incident radiation (q_{rad}) can be expressed as:

$$q_{rad} = \sigma\varepsilon_g(T_g^4 - T_{fab}^4) - \sigma\varepsilon_{fab}F_{fab-amb}(1-\varepsilon_g)(T_{fab}^4 - T_{amb}^4) \tag{2}$$

where σ is the Stefan-Boltzmann constant; ε_g, and ε_{fab} are the emissivity of the hot gases and the fabric, respectively; T_g, T_{fab}, and T_{amb} are the temperatures of the hot gases, the outside surface of the fabric, and the ambient air, respectively; and $F_{fab\text{-}amb}$ is the view factor which accounts for the geometry of the fabric in relation to the ambient air.

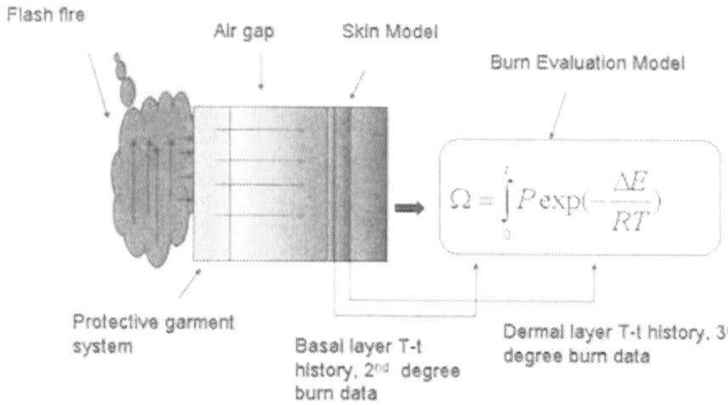

FIG. 2—*Elements of fabric air-gap skin model.*

The experimental foundation for the model was provided by three distinct experimental studies. The first experimental study characterized the thermal boundary conditions associated with the complex fire environment surrounding the manikin clothed in test garments. The second experimental base provided a means of estimating changes in the thermal physical properties of protective clothing material resulting from intense heat exposure. The third experimental analysis quantified the insulating air layers in skin-clothing systems and the changes in air gap distribution over the clothed manikin body due to garment shrinkage.

Flame temperatures measured over Pyroman® in the fire chamber ranged from 800–1400°C in a 4 s exposure to 2.0 cal/cm^2·s heat. The intensity of the incident thermal energy is normally distributed over the Pyroman® body. Variations in the incident heat flux result from the three-dimensional shape of the manikin surface and the complex and dynamic nature of the flame column surrounding the manikin. In order to compute heat transfer from the fire to the manikin, an overall heat transfer coefficient was estimated at each of the 122 thermal sensor locations [7].

Fabric thermal conductivity and heat capacity are the main factors controlling heat transfer in fabrics. The manikin model used a parameter estimation approach to quantify changes in fabric properties that occur as a result of intense heat exposure [8]. Experiments were conducted to estimate heat induced changes in the thermal conductivity and volumetric heat capacity of Kevlar®/PBI and Nomex®IIIA fabrics in short duration (4 s) intense fire exposures. The results show that both thermal conductivity and volumetric heat capacity decrease during the exposure [9].

The air trapped between the protective garment and the body is a major factor determining thermal protective insulation. Three-dimensional body scanning technology was used to measure the air gap distribution that exists between different size protective garments and the manikin body [9]. The process used data which were taken from a dressed manikin superimposed on data taken from an unclothed manikin.

The skin can be modeled as a three-layer structure composed of an avascular epidermis, a highly-perfused dermis, and a poorly-perfused subcutaneous tissue. In the three-layer structure skin model, different thermal properties are used in epidermis, dermis, and subcutaneous tissue [11]. Within each layer the tissue properties are assumed to be homogeneous, isotropic, and of uniform thickness. Pennes' Model was adopted to describe heat transfer in human skin. The bio-heat transfer equation [10] is written as:

$$\rho_s c_{p_s} \frac{\partial T}{\partial t} = \nabla \cdot (k_s \nabla T) + (\rho c_p)_b \omega_b (T_a - T) \tag{3}$$

where ρ_s and $c_{p,s}$ are the density and specific heat of human tissue; k_s is the thermal conductivity of human skin; ρ_b and $c_{p,b}$ are the density and the specific heat of the blood; ω_b is blood perfusion factor (0.00125 $m^3/s/m^3$); and T_a is arterial blood temperature.

The numerical model was validated using a Pyroman® manikin dressed in Kevlar®/PBI and Nomex®IIIA coveralls with exposure times from 3–4 s (Fig. 3). Results from the numerical model successfully correlate with experimental Pyroman® results for one-layer Kevlar®/PBI and Nomex®IIIA protective coveralls exposed for short intervals [9].

Numerical Model Predictions

Simulated Flash Fire in Manikin Chamber

The Pyroman® flash fire exposure was characterized by measuring the flame temperature and corresponding heat flux along each of the 122 sensors distributed over the manikin body. Figure 4 shows the distribution of heat flux values of the 122 sensors in Pyroman®, illustrating a bell-shaped distribution. In order to access this distribution normality further, a normal scores plot was performed, which demonstrated that the heat flux distribution with an average of 2.0 $cal/cm^2 \cdot s$ during a 4 s exposure exhibits an approximate normal distribution. Variations in heat fluxes are expected due to the three-dimensional shape of the manikin surface and because of the complex and dynamic nature of the flame column surrounding the manikin. Therefore, depending on the location on the manikin body (e.g., arms, legs, shoulders), sensors located in different positions on the manikin body will have different heat flux values with respect to the flash fire.

In the manikin test, an average heat flux of 2.0 $cal/cm^2 \cdot s$ measured from the 122 sensors on the manikin is used to simulate flash fire conditions, and the standard deviation of these heat flux distributions is between 0.25–0.5. Figure 5 shows the bell-shaped distributions with a different standard deviation. The larger standard deviation indicates that more extreme high and low flux values occurred during the exposure. In order to examine the effects of different fire distributions, a series of heat flux data with different distributions was generated to simulate statistically the different fire distributions. These data, as a model fire boundary input, were predicted with a one-layer garment using the numerical model. The model is based on one-layer Kevlar®/PBI coverall. The major parameters used in the numerical model are described in Table 1. Figure 6 outlines the model calculations.

The standard deviation of the heat fluxes of all 122 Pyroman® sensors predicted by this model does not show a significant influence on third-degree burn prediction. The second-degree burn prediction, however, decreases as the standard deviation increases, as illustrated in Fig. 6. This is expected; as the standard deviation increases, more low heat flux values are generated.

This lowered heat flux value could help reduce the burn predictions. For 3 s of exposure, no significant difference was predicted on the results of second-degree burn.

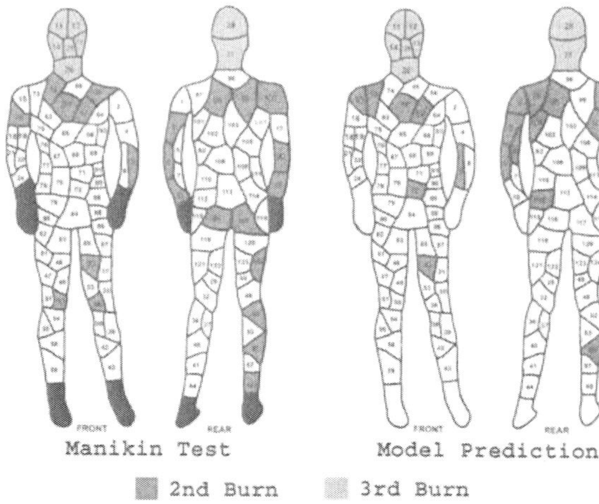

FIG. 3—*A comparison of burn distribution from manikin test and model prediction for 3 s exposures.*

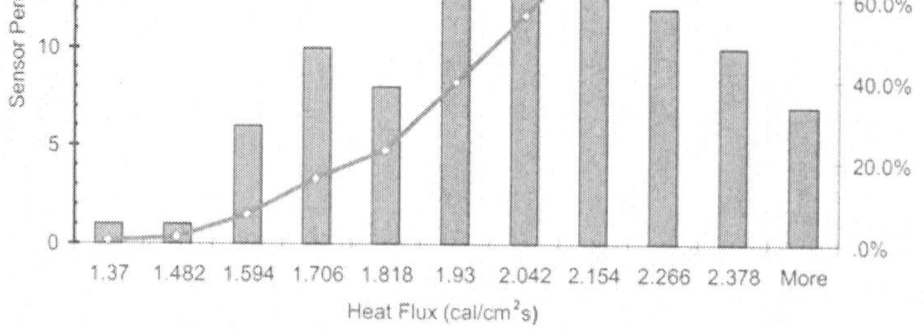

FIG. 4—*Histogram and cumulative curve of heat fluxes of 122 sensors over the manikin in 4 s exposures.*

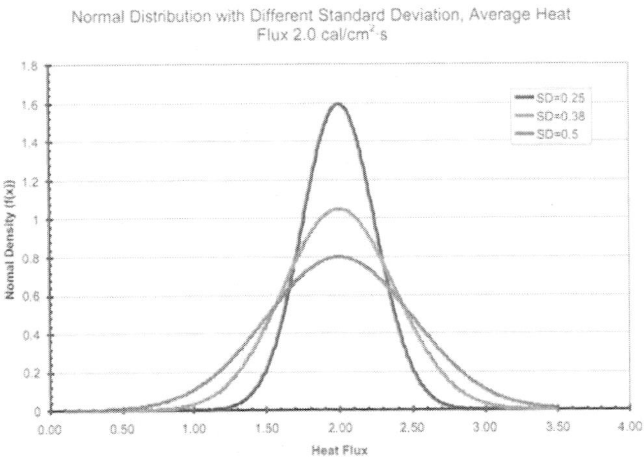

FIG. 5—*Heat flux distribution with different standard deviation.*

TABLE 1—*Parameters used in numerical model.*

Model Parameters	Value
Average Heat Flux	2.00 cal/cm^2·s
Fabric Weight	153g/m^2
Garment Air Gap Sizes	Kevlar®/PBI Coverall size 42 [9]
Garment Size	42
Burning Time (s)	3 or 4

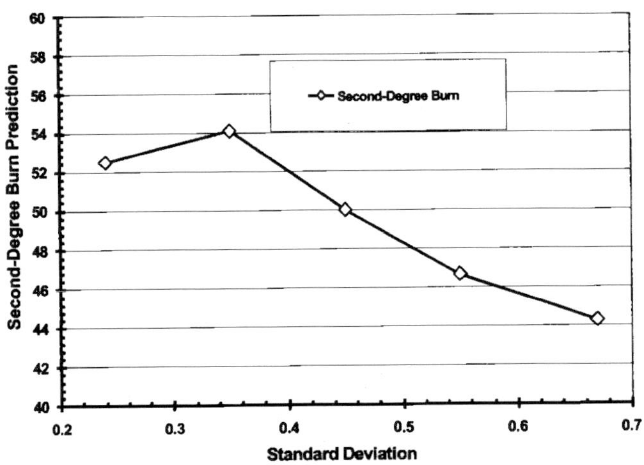

FIG. 6—*One-layer coverall manikin fire testing predictions with different fire distribution (overall average heat flux is 2.0 cal/cm^2·s).*

Skin Model

The variations in the skin model were analyzed using the manikin numerical model based on a one-layer coverall (Table 1). Table 2 summarizes the thermal physical properties used in one-layer and three-layer skin models [11].

Skin Temperature Distribution—One of the functions of human skin is to help regulate the body's core temperature. The core temperature of the body must be maintained within a small range around 37°C in order to keep biochemical reactions proceeding at required rates. Under normal ambient conditions, the skin surface temperature is about 32.5°C–34°C. Some skin models proposed constant initial skin temperatures of 32.5°C [12], 34°C [13], or 37°C [14]. Some other models suggest linear initial or higher order initial temperature distribution. For a given incident heat flux history transmitted through the garment under a flash fire condition, different initial temperature distribution in a skin model could affect the temperature rising rate at specific depth in the skin. In this study, five initial temperature distributions are selected. The first three are constant at 32.5, 34, and 37°C from surface to subcutaneous base; the fourth is a linear distribution from 32.5°C surface to 37°C subcutaneous; and the fifth is a quadratic distribution from 32.5°C surface to 37°C subcutaneous base. Temperature distributions in the different skin models are also examined. The one-layer skin model was constructed using properties shown in Table 2. The model prediction results are illustrated in Tables 3 and 4.

TABLE 2—*Thermal properties in different skin models.*

Properties	One-Layer	Three-layer [11]		
	Tissue	Epidermis	Dermis	Subcutaneous
Thermal Conductivity (W/m°C)	0.335	0.225	0.523	0.167
Volumetric Heat Capacity ($J/m^3 \cdot K*10^6$)	3.87	4.32	3.87	2.76

TABLE 3—*Temperature distribution effects in three-layer skin model in a 4 s exposure.*

Burn Prediction	Constant at 32.5°C	Constant at 34°C	Constant at 37°C	Linear from 32.5 to 37°C	Quadratic from 32.5 to 37°C
Second %	43.44	52.46	77.05	44.26	49.18
Third %	7.38	8.2	9.84	7.38	8.2
Total (%)	50.82	60.66	86.89	51.64	57.38

TABLE 4—*Temperature distribution effects in one-layer skin model in a 4 s exposure.*

Burn Prediction	Constant at 32.5°C	Constant at 34°C	Constant at 37°C	Linear from 32.5 to 37°C	Quadratic from 32.5 to 37°C
Second %	73.77	74.5	73.77	74.59	74.59
Third %	6.56	6.56	7.38	6.56	6.56
Total (%)	80.33	81.06	81.15	81.15	81.15

In the three-layer skin model, the predictions are very sensitive to initial temperature distributions. The linear and quadratic temperature distributions show almost a 10 % difference in burn predictions, while the constant temperature (37°C) distribution compared to the linear and quadratic indicates a large influence on second body burn prediction. The changes of these temperature distributions show little influence on third-degree burns. This result is expected because the temperature difference of these three distributions is very small at the interface of dermis and subcutaneous. In the one-layer skin model, however, initial temperature distribution shows little effect on burn predictions. The simulation results demonstrate that the burn predictions on initial temperature distribution in the one-layer skin model are not as sensitive as in the three-layer skin model. This result is attributable to the relative large thermal conductivity used in the one-layer skin model.

Blood Perfusion—A simulation was performed to determine the influence of the blood perfusion in the skin model on burn predictions in flash fire conditions. Two different blood perfusion values [9] were used in the skin model with comparison to the model without blood perfusion. The numerical results lead us to believe that blood perfusion in the skin model shows no effect on burn predictions in short duration flash fire conditions. Blood perfusion does not practically influence the burn prediction in terms of second and third body burns for the short exposure. However, as shown in Table 6, it changes the time to second- or third-degree burn at some specific individual sensor locations, especially when the prediction time is relatively longer. The time change is anticipated because the longer time allows the blood an opportunity to carry away some of the energy before damage is sustained.

TABLE 5—*Blood perfusion influence on burn predictions.*

	G = 0 ($m^3/s/m^3$)		G = 1.25×10^{-3} ($m^3/s/m^3$)		G = 2.5×10^{-3} ($m^3/s/m^3$)	
Sensor No.	Time to Second Burn (s)	Time to Third Burn (s)	Time to Second Burn (s)	Time to Third Burn (s)	Time to Second Burn (s)	Time to Third Burn (s)
3	2.27	12.3	2.27	12.45	2.27	12.55
20	2.62	14.2	2.62	14.45	2.62	14.75
85	6.61	no	6.62	no	6.63	no
92	7.16	no	7.19	no	7.61	no

Conclusion

The results of the numerical model predictions demonstrate that a different fire distribution in manikin fire testing exhibits influence on second burn predictions for a one-layer garment in a 4 s exposure. For a 3 s exposure, however, no significant difference is found on burn predictions. In addition, a decreased trend of prediction is found when the standard deviation increases. The results suggest that both the overall average heat flux and the precise standard deviation of a simulated flash fire should be indicated in order to enhance the manikin prediction.

Assuming blood perfusion in a skin model has minimal effect on predicted body burns in these exposures. The model assuming blood perfusion slightly increases the time to second- or third-degree burns, especially when the time to get second- or third-degree burns exceeds 5 s.

Different temperature distributions in the skin models demonstrate a large effect on burn predictions in the three-layer skin model. The effect is pronounced in the constant temperature distribution compared to linear and quadratic. In the one-layer skin model, however, initial temperature distribution shows little effect on burn predictions. Large second-degree burn predictions are observed in the one-layer skin model relative to the three-layer skin model. This study suggests that the use of different skin models and their initial temperature distributions can greatly affect burn predictions of garment testing in an instrumented manikin. Therefore, a precise skin model selection and its standardization would be beneficial for manikin testing in thermal protective performance evaluation.

References

[1] Pawar, M., and Barker, R. L., "Analyzing the Performance of Protective Garments in Manikin Fire Tests," *Proceedings of the Fifth Scandinavian Symposium on Protective Clothing*, Elsinore, Denmark, 1997.

[2] Henriques, F. C. and Moritz, A. R., "Studies of Thermal Injury: The Conduction of Heat to and Through Skin and the Temperatures Attained Therein. A Theoretical and Experimental Investigation," *American Journal of Pathology*, Vol. 23, 1947, p. 531.

[3] Buettner, K., "Effects of Extreme Heat and Cold on Human Skin II. Surface Temperature, Pain and Heat Conductivity in Experiments with Radiant Heat," *Journal of Applied Physiology*, Vol. 3, 1951, pp. 703–713.

[4] Mehta, A. K., and Wong, F., "Measurement of Flammability and Burn Potential of Fabrics," *Summary Report from Fuels Research Laboratory*, Massachusetts Institute of Technology, Cambridge, Massachusetts, 1972.

[5] Weaver, J. A., and Stoll, A. M., "Mathematical Model of Skin Exposed to Thermal Radiation," *Aerospace Medicine*, Vol. 40, 1969, pp. 24–30.

[6] Torvi, D. A., "Heat Transfer in Thin Fibrous Materials Under High Heat Flux Conditions," Doctoral Thesis, University of Alberta, Canada, 1997.

[7] Beck, J. V., Blackwell, B., and St. Clair, C. R., Jr., "Inverse Heat Conduction: Ill-Posed Problems," John Wiley & Sons, Inc., New York, 1985.

[8] Beck, J. V. and Arnold, K., "Parameter Estimation in Engineering and Science," Wiley, New York, 1977.

[9] Song, G., "Modeling Thermal Protection Outfits for Fire Exposures," Doctoral Dissertation, North Carolina State University, Raleigh, NC, 2002.

[10] Pennes, H. H., "Analysis of Tissue and Arterial Blood Temperatures in Resting Human Forearm," *Journal of Applied Physiology*, Vol. 1, 1948, pp. 93–122.

[11] ASTM F 1930-00: Standard Test Method for Evaluation of Flame Resistant Clothing for Protection Against Flash Fire Simulations Using an Instrumented Manikin, *Annual Book of ASTM Standards*, Vol. 11, No. 3, ASTM International, West Conshohocken, PA, 2003.

[12] Stoll, A. M., and Greene, L. C., "Relationship Between Pain and Tissue Damage Due to Thermal Radiation," *Journal of Applied Physiology*, Vol. 14, 1959, pp. 373–382.

[13] Diller, K. R., and Hayes, I. J., "A Finite Element Model of Burn Injury in Blood Perfused Skin," *Transactions of the ASME, Journal of Biomechanical Engineering*, Vol. 105, 1983, pp. 300–307.

[14] Henriques, F. C., Jr., "Studies of Thermal Injuries V. The Predictability and the Significance of Thermally Induced Rate Process Leading to Irreversible Epidermal Injury," *Archives of Pathology*, Vol. 43, 1947, pp. 489–502.

Journal of ASTM International, February 2005, Vol. 2, No. 2
Paper ID JAI12118
Available online at www.astm.org

Carole A. Winterhalter,[1] Rosemary A. Lomba,[1] David W. Tucker,[2] and O. David Martin[3]

Novel Approach to Soldier Flame Protection

ABSTRACT: The U.S. Army desires low cost combat uniform fabrics that provide flame protection, visual and near infrared camouflage, comfort, and durability. Prior to material development, the flame threat and hazard was investigated and characterized. Flammability related test methods were reviewed and pass/fail criteria for instrumented manikin testing was established based on military medical doctrine. Various fibers, fiber blends, and functional finishes were investigated and novel blend materials were developed. Textile manufacturing methods not traditionally used in military protective clothing such as core-spun yarn spinning and direct fiber-to-fabric non-woven technologies were also investigated. When compared to the standard issue combat uniform, the best performing protective clothing configuration demonstrated a reduction in total body burn from 88 % to 8 %. Safe exposure limits of up to 5 s were established.

KEYWORDS: flame protective clothing systems, flame resistant textiles, combat uniform fabrics, military, army, soldier

Introduction

U.S. Army tankers and aviators of all services are required to wear flame resistant clothing systems made from Nomex®[4] and Kevlar®[5] fiber blend fabrics. The infantry, which wears a nylon and cotton blend fabric, desires flame protection, but uniforms made from Nomex and Kevlar blend fabrics are too expensive to issue to every soldier. The objective of this effort is to develop new materials and system protection strategies that are affordable and improve the survivability of the individual soldier. Prior to material development, the flame threat and hazard was investigated and characterized. Flammability related test methods were reviewed and pass/fail criteria for manikin testing were established based on military medical doctrine. In addition to low cost flame protection, a fabric is desired that provides the multi-functionality of the all-purpose combat uniform currently in use, i.e., visual and near-infrared camouflage protection, comfort in environmental extremes, and durability and protection against the elements. Fibers and fabric blends investigated include: melamine (Basofil®[6]), flame retardant-treated (FRT) cotton, FRT lyocell (Tencel®[7]) rayon, FRT cotton/Kevlar/nylon, carbonized rayon/Nomex, Kevlar/FR rayon, and PBI®[8]/FRT cotton.

Manuscript received 18 November 2003; accepted for publication 28 July 2004; published February 2005.
Presented at ASTM Symposium on Performance of Protective Clothing: Global Needs and Emerging Markets: 8th Symposium on 13 January 2004 in Tampa, FL.
[1] Textile Technologist, U.S. Army Research, Development and Engineering Center, Natick, MA 01760-5019.
[2] Senior Operations Research Systems Analyst, U.S. Army Research, Development and Engineering Center, Natick, MA 01760-5020.
[3] Senior Research Chemist, DuPont Advanced Fibers Systems, Richmond, VA 23234.
[4] Nomex® is a trademark of E.I. duPont de Nemours & Company, Wilmington, DE.
[5] Kevlar® is a trademark of E.I. duPont deNemours & Company, Wilmington, DE.
[6] Basofil® is a trademark of BASF, Germany.
[7] Tencel® is a trademark of Courtaulds Fibres Limited, England.
[8] PBI® is a trademark of CNA Holdings Inc., Summit, New Jersey.

Flame Threat and Hazard

Prior to material development, the flame threat and hazard was investigated and characterized. The likelihood of the military flame burn hazard on the battlefield can be attributed to three components: threat-generated flame burns, incidental or secondary flame burns, and burn accidents. The threat-generated burn hazard results from the direct employment of a flame and thermal weapon. Incidental or secondary flame burn hazards result from flame and thermal weapons or other threats (i.e., ballistic, blast, chemical, directed energy) igniting battlefield combustibles, including clothing or equipment that can also present a burn hazard, in addition to the direct effect of the specific threat. Accidents comprise the remaining burn hazard and account for over half of all burns. Historically, even in combat, flame burn injuries more often result from the ignition of battlefield combustibles, due to accidents or secondary flame effects in the environment, than from direct flame and thermal threats [1]. The thermal characteristics of many battlefield combustibles were reviewed. According to Kim, the estimated values of the heat flux of thermobaric, incendiary, and flame weapons, and JP-8 fuel converge to a common heat flux of about 2.0 cal/(cm^2 s) over time [2]. Since the likelihood of survival of a direct hit from a flame and thermal weapon is low based on the initial ignition temperatures and known blast effects, and the fact that most burn injuries result from the secondary effects of the environment, a thermal flux of 2.0 cal/(cm^2 s) was selected because it is representative of a reasonable military fire hazard that is survivable if exposure time is short. Appropriately developed protective clothing should provide critical seconds of increased escape time to distance oneself from the flame hazard and thereby provide a lesser, survivable exposure time.

Clothing Systems Test and Evaluation

Flammability test methods were reviewed and the instrumented manikin test, ASTM F 1930, which targets a thermal flux of 2.0 cal/(cm^2 s), was selected for the clothing system test and evaluation. The establishment of pass/fail criteria for the test was based on military medical doctrine. A partial listing of the accepted criteria for transferring patients to a burn center include: (a) partial-thickness burns of 20 % of total body surface area burned (TBSAB) or greater in adults, (b) full-thickness burns exceeding 5 % TBSAB, and (c) burns involving the face, feet, hands, perineum, or major joints. During armed conflict, however, additional criteria include casualties with small burns involving the hands, feet, or perineum that may be considered for early evacuation since their activity will be disproportionately limited [3]. In addition, during combat, available resources must be expended on those individuals with the greatest chance for survival. Casualties at the extremes of age (that is, those under 10 or over 60 years old) or those with 70 % of the total body surface area burned will, in general, do poorly, and should be allocated a lesser share of the resources. The care of casualties with less than 20 % TBSAB can safely be delayed pending either their evacuation to a higher-echelon medical treatment facility or the availability of more resources [4,5]. Pass/fail criteria for the instrumented manikin test were established as no more than a 20 % TBSAB based on the likelihood of the availability of treatment during combat.

Instrumented manikin testing was conducted on both aviator and tanker protective clothing systems beginning with summer weight and adding clothing layers up to the full winter weight configuration (Table 1). The clothing systems were tested from 3 to 10 s of exposure (Table 2) [6]. Safe exposure limits were established for each clothing configuration (Table 3). Safe exposure limits were identified based on the protection provided by each system. This approach was selected for two reasons: 1) many military protective clothing systems provide protection against multiple battlefield threats, hazards, and conditions, and the protection mechanisms for each may be mutually exclusive, and 2) the flame and thermal exposure time is random and therefore difficult to specify.

Exposure limits increase with each additional clothing layer. The first line of defense against the flame assault is the outer-layer, which in every configuration is flame and ignition resistant. Each additional clothing underlayer adds insulation and increases protection time. Tests performed with both cotton long underwear and Nomex long underwear show no difference in performance. These findings corroborate other studies [7] demonstrating that the entire clothing system, especially the underlayers, does not necessarily need to be made from flame resistant materials. The aviator and tanker systems performed similarly at various exposures with the exception of the summer weight configuration. The tanker coverall demonstrated a second-degree body burn of 8 % versus 18 % for the two-piece aviator system. The differences are likely due to differences in material type and garment construction. The tanker coverall was made from a 4.5 oz/yd^2 producer colored Nomex fabric, and the aviator coat and trousers were made of a 5.5 oz/yd^2 camouflage printed Nomex and Kevlar fiber blend.

TABLE 1—*Aviator and tanker clothing configurations.*

Configuration	A. Aviator	B. Tanker
1	T-shirt [8] and Briefs [9]	T-shirt and Briefs
		Combat Vehicle Crewman (CVC)
	Aircrew Coat [10] and Trouser[11]	Coverall [12]
	Lightweight Balaclava [13]	Lightweight Balaclava
	Nomex Gloves [14]	Nomex Gloves
	Combat Boots [15]	Combat Boots
2	Cotton Long Underwear	Cotton Long Underwear
	Aircrew Coat and Trouser	Combat Vehicle Crewman Coverall
	Lightweight Balaclava	Lightweight Balaclava
	Nomex Gloves	Nomex Gloves
	Combat Boots	Combat Boots
3	Nomex Long Underwear [16]	Nomex Long Underwear
	Aircrew Coat and Trouser	Combat Vehicle Crewman Coverall
	Lightweight Balaclava	Lightweight Balaclava
	Nomex Glove	Nomex Gloves
	Combat Boots	Combat Boots
4	Nomex Long Underwear	Nomex Long Underwear
	Aircrew Coat and Trouser	Combat Vehicle Crewman Coverall
	Aircrew Jacket and Liner [17]	Combat Vehicle Crewman Jacket [18]
	Lightweight Balaclava	Lightweight Balaclava
	Nomex Gloves	Nomex Gloves
	Combat Boots	Combat Boots
5	Nomex Long Underwear	Nomex Long Underwear
	Aircrew Coat and Trouser	Combat Vehicle Crewman Coverall
	Bib Overall [19]	Bib Overall
	Aircrew Jacket and Liner	Combat Vehicle Crewman Jacket
	Cold Weather Balaclava [20]	Cold Weather Balaclava
	Nomex Gloves	Nomex Gloves
	Combat Boots	Combat Boots

TABLE 2—*Summary of burn injury prediction.*

	Exposure Time											
	3 s % Predicted Burn*			4 s % Predicted Burn*			6 s % Predicted Burn*			10 s % Predicted Burn*		
Ensemble	2nd degree	3rd degree	Total	2nd degree	3rd degree	Total	2nd degree	3rd degree	Total	2nd degree	3rd degree	Total
1A	18	2	20	23	7	30	25[a]	31[a]	56[a]			
2A	0	2	2	2	2	4	17	3	19			
3A	0	2	2	2	2	4	26	2	28			
4A				2	2	4	9	2	11			
5A				0[a]	2[a]	2[a]	0	2	2	1	4	5
1B	8	2	10	13	16	29	22[a]	43[a]	65[a]			
2B	1	2	3	3	3	6	18	14	32			
3B	0	2	2	2	2	4	21	9	30			
4B				2	2	4	8	4	13			
5B				0[a]	2[a]	2[a]	0	2	2	2	3	5

* Average of replicate measurements.
[a] Result of one measurement.

TABLE 3—*Safe exposure limits for military flame resistant clothing.*

Configuration	1	2	3	4
Layer				
One	T-shirt, Briefs Cotton	Long Underwear Cotton or Nomex	Long Underwear Nomex	Long Underwear Nomex
Two	Coverall or ABDU Nomex	Coverall or ABDU Nomex	Coverall or ABDU Nomex	Coverall or ABDU Nomex
Three	...	...	Jacket Nomex	Bib-Overalls Nomex
Four	...	...	...	Jacket Nomex
% Body Burn	Less than 20 % At 3 s	Less than 20 % At 4 s	Less than 20 % At 6 s	Less than 20 % At 10 s

Materials Development and Discussion

The overall objective was to develop a low cost flame resistant fabric for use by the infantry that also provides the multi-functionality of the all-purpose combat uniform fabric currently in use, i.e., visual and near-infrared camouflage, comfort in climatic extremes such as hot and humid, and hot and dry, and durability and protection against the elements. As a guide, the military standard camouflage printed Nomex[9] fabric was used for general performance goals. Fibers, fiber blends, and functional finishes investigated include: Basofil, FRT cotton, FRT cotton/nylon, FRT Tencel, FRT cotton/Kevlar/nylon, carbonized rayon/Nomex, Kevlar/FR rayon, Nomex/FR rayon, PBI, and PBI/FRT cotton. Weight and strength data for these fabrics are listed (Table 4) with those for the military camouflage printed Nomex fabric.

Many of the inherently flame resistant fibers were eliminated for use in a homogeneous fabric due to their high cost and the requirement for visual and near infrared camouflage. The high polymer orientation of the aramids and PBI, for example, contributes to their flame resistance, but also reduces or eliminates their ability to be dyed with traditional dyestuffs due to the lack of chemical dye sites. Some of these materials may achieve coloration by pigment

[9] MIL-C-83429, Cloth, Plain and Basket Weave, Aramid.

injection in solution form, but their versatility is limited. Aramid blends are dyed and camouflage-printed using proprietary technology that adds significantly to the final cost of the finished fabric. Still prized for their inherent flame resistance, some of these fibers were blended with low cost fibers to enhance the overall flame resistance of the fabric. Flame retardant rayon, which is inherently flame resistant rather than flame retardant treated, was blended with the aramids in 60/40 and 35/65 % blend ratios, but these materials fell short of the desired fabric strength and the camouflage print demonstrated poor colorfastness.

TABLE 4—*Performance characteristics of developmental and commercial materials.*

Fiber Blend	Weight, oz/sq.yd	Tearing Strength, lbs	Breaking Strength, lbs	Flame Resistance
MIL-C-83429 Class 5	4.7–6.0	9 × 7 (min)	180 × 100 (min)	Yes
92 % Nomex 5 % Kevlar 3 % P140				
Class 6	4.3–5.0	12 × 8 (min)	180 × 100 (min)	Yes
93 % Nomex 5 % Kevlar 2 % P140 (Spun Laced)	3.1	Does not tear	109 × 100	Yes
58 % FR Cotton (sheath) 2 % Mannacryl (sheath) 40 % Kevlar (core)	4.5	8 × 8	107 × 61	Yes
80 % FR Cotton 20 % PBI	5.0	9 × 7	83 × 48	Yes
58 % FR Cotton 27 % Kevlar 15 % Nylon	6.3	3 × 3	153 × 160	Yes
58 % FR Rayon 40 % Kevlar 2 % Conductive Fiber	6.4	10 × 13	150 × 100	Yes
65 % Nomex 35 % Rayon	6.5	9 × 6	131 × 102	Yes
80 % FR Cotton 20 % PBI	6.6	11 × 6	139 × 75	Yes
Carbonized Rayon Nomex	6.9	10 × 8	198 × 125	Yes
100 % FR Lyocell (Tencel)	7.4	6 × 7	180 × 117	Yes
88 % FR Cotton 12 % Nylon	7.7	6 × 7	99 × 72	Yes
88 % FR Cotton 12 % Nylon	11.0	5 × 9	161 × 154	Yes

Flame retardant-treated cotton has long been the industry standard for use in low cost flame resistant industrial work wear. However, the most commonly used treatment, Indura®[10], adds 20 % to the weight of the fabric. Flame-retardant treated cotton was blended with nylon in 88/12 % blend ratios, where the nylon was added to improve strength. While the addition of the nylon did not negatively impact the flame resistance, a heavier weight fabric of 11 oz/yd^2 was required to achieve relatively acceptable breaking and tearing strengths.

Flame retardant-treated cotton was also blended with Kevlar and nylon to enhance flame resistance and improve abrasion resistance in a 58/40/15 % blend ratio. While the Kevlar was

[10] Indura is a trademark of Westex, Inc., Chicago, IL.

the strongest of the three fibers, it occupied less than 50 % of the total material composition to keep costs down. The strength of the fabric is dictated by the dominant fiber, which in this case was the lower strength cotton. In addition, the high end and pick count required to anchor the Kevlar fiber detrimentally reduced the fabric tearing strength to 3 lbs in the warp and filling directions. PBI was blended with flame retardant-treated cotton in a 20/80 % blend ratio in 5.0 and 6.6 oz/yd^2 weights; however, the performance characteristics of the predominate fiber, cotton, prevailed.

Basofil fiber demonstrated low fiber tenacity and developmental efforts were directed toward insulation, knitted headwear, and hand wear applications where high strength was not a critical factor. Blends of carbonized rayon and Nomex were investigated and while they demonstrated good strength performance they could not be dyed and camouflage-printed. Flame retardant-treated Tencel demonstrated good strength but the camouflage demonstrated poor colorfastness performance [21]. Core spun yarns were investigated and developed with the primary intent of manufacturing a yarn with a high strength, inherently flame resistant core, and a low cost readily camouflage-printable sheath fiber. The best-performing material combination was a cotton sheath, Kevlar core yarn. These materials also fell short on strength because only the Kevlar-based core and not the sheath contributed to the fabric strength.

While all of the developmental materials investigated met the fabric bench scale flame performance goals (ASTM D 6413; 2.0 s, maximum after flame; 25.0 s, maximum after glow; 4.0 in. maximum char length), strength and other performance requirements such as colorfastness of the camouflage fell short. As an alternative, we investigated a novel approach to soldier flame protection and changed the focus of fabric development from flame retardant-treated wovens made from low cost fiber blend yarns to direct fiber to fabric manufacturing, potentially saving on yarn spinning, fabric weaving, and finishing. A spun-laced non-woven Nomex and Kevlar blend fabric developed by Dupont demonstrated strength that was equal to or greater than the former all-cotton Hot Weather Battledress Uniform (BDU) [22,23] and half the weight. Due to its lightweight, open, air permeable construction it can be worn over the existing BDU and together this configuration provides a 40 % cost savings over the existing camouflage-printed Nomex/Kevlar Aircrew BDU used by Army aviators. Instrumented manikin testing (Table 5) demonstrated that when the Nomex coverall (Nomex® Limitedwear[11]) was worn over the standard BDU the total body burn was reduced from 88 to 8 %. Safe exposure limits up to 5 s were established. The non-woven Nomex outer shell provides ignition resistance, and the underlayers of the BDU, t-shirt and briefs, provide thermal insulation.

TABLE 5—*Results of thermoman burn injury prediction.*

	3 Seconds			4 Seconds			5 Seconds			6 Seconds		
Ensemble	2nd	3rd	Total	2nd	3rd	Total	2nd	3rd	Total	2nd	3rd	Total
T-shirt, briefs,	9	71	80									
HWBDU	14	74	88									
T-shirt, briefs,	1	6	7	4	7	11	4	11	15	10	14	24
HWBDU,	0	7	7	5	7	12	3	9	12	7	13	20
Limitedwear	1	7	8	3	8	11	7	11	18	8	13	21

[11] Nomex® Limitedwear is a trademark of E.I. duPont deNemours & Company, Wilmington, DE.

Conclusions and Recommendations

Flame burn injuries more often result from the ignition of battlefield combustibles than from direct flame and thermal threats. A review of the estimated heat flux of several battlefield combustibles shows that they converge to 2.0 cal/(cm^2 s) after a few minutes. This thermal flux was selected as representative of a reasonable military fire hazard that is survivable if the exposure is short and protective clothing is worn. A pass/fail criterion for clothing systems flammability testing was selected as no greater than a 20 % body burn based on military medical doctrine. Material selection and clothing configuration influence system flame protection. The entire configuration does not necessarily need to be flame resistant; however, the outer-most materials are the first line of defense against flame and thermal assault and must be flame resistant. Each additional clothing underlayer adds insulation and increases protection time. Military flame protective clothing systems provide 3 to 10 s of protection depending on the number of clothing layers worn. The Air Crew BDU and CVC coveralls provide 3 s of protection. The addition of long underwear (cotton or Nomex) increases protection time to 4 s. The addition of a bib overall and jacket provide 10 s of protection. The Infantry, which wears a uniform made from a nylon/cotton blend fabric, desires low cost flame protection. Various fibers, fiber blends, and functional finishes were investigated and novel blend materials were developed. While a direct fabric substitution could not be made, a supplemental flame resistant overgarment is recommended. When compared to the standard issue combat uniform, the combination of the HWBDU and Nomex coverall demonstrated a reduction in total body burn from 88 to 8 %. Safe exposure limits of up to 5 s were established.

References

[1] Tucker, D. W., Rei, S. A., *Soldier Flame/Thermal Hazard Assessment, Natick/TR-99/039L,* U.S. Army Soldier and Biological Chemical Command, Natick, MA, 1999.

[2] Kim, I. L., *Battlefield Flame/Thermal Threats or Hazards and Thermal Performance Criteria, TR-00/015L,* U.S. Army Soldier and Biological Chemical Command, Natick, MA, 2000.

[3] Committee on Trauma, "Resources for Optimal Care of Patients with Burn Injury," *Resources for Optimal Care of the Injured Patient,* American College of Surgeons, Chicago, 1990, pp. 57–60.

[4] *The Military Medicine Series, Part I. Warfare, Weaponry, and the Casualty, Volume 5: Conventional Warfare: Ballistic, Blast, and Burn,* Office of the Surgeon General, Department of the Army, United States of America, Washington, DC, 1990.

[5] Chapter 3, Burn Injury, *Emergency War Surgery, 2nd U.S. Rev. of The Emergency War Surgery NATO Handbook,* U.S. Department of Defense, T. E. Bowen and R. F. Bellamy, Eds., Washington, DC, 1989, pp. 35–56.

[6] A Report on the Thermal Protective Performance (TPP) and the Pyroman Evaluation for the U.S. Army Soldier Systems Command, Center for Research on Textile Protection and Comfort (T-PACC), College of Textiles, North Carolina State University, Raleigh, NC, 1999.

[7] Capecci, T. M. and Swaverly, C. B., "Flammability and Thermal Protection Testing of Long Underwear for Navy and Marine Corps Aircrew Use," *Performance of Protective Clothing: Fifth Volume, ASTM STP 1237,* ASTM International, West Conshohocken, PA, 1995, pp.

625–634.
[8] MIL-U-44096, Undershirt, Cotton, Brown 436, Defense Support Center – Philadelphia, PA.
[9] MIL-D-43783, Brief Drawers, Cotton, Brown 436, Defense Support Center – Philadelphia, PA.
[10] MIL-C-44371, Coat, Aircrew, Combat, Defense Support Center – Philadelphia, PA.
[11] MIL-T-44372, Trousers, Aircrew, Combat, Defense Support Center – Philadelphia, PA.
[12] MIL-C-44077, Coveralls, Combat Vehicle Crewman's, Defense Support Center – Philadelphia, PA.
[13] MIL-H-24936, Hood, Anti-Flash, Flame Resistant, Defense Support Center – Philadelphia, PA.
[14] MIL-DTL-81188, Gloves, Flyer's Summer, Type GS/FRP-2, Defense Support Center – Philadelphia, PA.
[15] MIL-B-44152, Boots, Combat, Mildew and Water Resistant, DHS, Defense Support Center – Philadelphia, PA.
[16] MIL-D-85040, Drawers and Undershirts, Flyer's, Anti-Exposure, Aramid, High Temperature Resistant, CWU-43/P and CWU-44/P, Defense Support Center - Philadelphia, PA.
[17] MIL-DTL-31010, Jacket, Aircrew, Cold Weather, With Removable, Liner, and Hood, Defense Support Center – Philadelphia, PA.
[18] MIL-J-43924, Jacket, Cold Weather, High Temperature Resistant, Defense Support Center – Philadelphia, PA.
[19] MIL-O-44109, Overall, Combat Vehicle Crewman's, Defense Support Center – Philadelphia, PA.
[20] MIL-H-44265, Hood, Combat Vehicle Crewman's (Balaclava), Defense Support Center – Philadelphia, PA.
[21] Winterhalter, C. A., *Experimental Battledress Uniform Fabrics Made From Amine Oxide Solvent Spun Cellulosic Fibers, Natick/TR-02/007,* U.S. Army Soldier and Biological Chemical Command, Natick, MA, 2002.
[22] MIL-DTL-44048, Coats, Camouflage Pattern, Combat, Defense Support Center - Philadelphia, PA.
[23] MIL-T-44047, Trousers, Camouflage Pattern, Combat, Defense Support Center - Philadelphia, PA.

Author Index

Subject Index